Saudi Arabia and Oil Diplomacy

Sheikh Rustum Ali

The Praeger Special Studies program—utilizing the most modern and efficient book production techniques and a selective worldwide distribution network—makes available to the academic, government, and business communities significant, timely research in U.S. and international economic, social, and political development.

Saudi Arabia and Oil Diplomacy

PRAEGER SPECIAL STUDIES IN INTERNATIONAL POLITICS AND GOVERNMENT

Praeger Publishers New York Washington London

Library of Congress Cataloging in Publication Data

Ali, Sheikh Rustum.
Saudi Arabia and oil diplomacy.

(Praeger special studies in international politics and government)
Bibliography: p.
Includes index.
1. Petroleum industry and trade—Saudi Arabia.
2. Saudi Arabia—Foreign relations—United States.
3. United States—Foreign relations—Saudi Arabia.
4. Jewish-Arab relations—1973- I. Title.
HD9576.S33A44 327.53'8'073 75-36413
ISBN 0-275-56260-3

PRAEGER PUBLISHERS
111 Fourth Avenue, New York, N.Y. 10003, U.S.A.

Published in the United States of America in 1976
by Praeger Publishers, Inc.

Printed in the United States of America

To the memory of my father,
whose interest in advanced study
made my childhood dream a reality

PREFACE

This study seeks to analyze the use of oil by Saudi Arabia as a weapon of diplomacy in the settlement of the Arab-Israeli conflict within a framework that highlights the salient variables such as politics, economics, international relations, and diplomacy. Saudi Arabia wanted to use its oil as an instrument to bring pressure on the United States to moderate its pro-Israeli policy in the Middle East. The external need for Saudi oil resources was not sufficiently exploited politically until the beginning of the oil embargo in 1973. Till that time, despite the difficulty of separating economics from politics, the Arabs tried to keep the two apart as much as possible, for fear of a possible economic loss.

It is our hypothesis that oil, a depletable but replaceable commodity, cannot be used as a weapon of diplomacy in the long run. In the short run, its use as a diplomatic weapon in the settlement of the Arab-Israeli conflict has a limited scope. The hypothesis will be tested by examining the impact of the Arab pressure on the United States through the 1973-74 oil embargo for the settlement of the Arab-Israeli conflict and by studying the trend in the development of the alternative sources of energy.

The increasing demand for oil in the developed as well as developing countries has caused oil to become the most important source of energy in the world. International oil is a multibillion dollar industry that affects, in varying degrees, the balance of payments of oil exporting and consuming countries. Because of the embargo of oil by Saudi Arabia and other Arab nations, oil has inevitably been enmeshed into international political and economic relations and diplomatic activity.

The stated aim of this study is to examine the relationship between oil and diplomacy through a case study of Saudi Arabia. It may be assumed that, while oil demand will grow in the following decades, further oil demand until the end of this century would depend significantly on technological developments in energy production and consumption. Energy resources are abundant. Given the technology and money to develop them, it can be visualized that by the end of the century the demand for oil will be sharply curtailed, thus depriving it of its present significance and of its role as a weapon of diplomacy.

The answer to the alternative sources of oil is sought in energy production and consumption, curtailment or competition. This study evaluates the assumption that oil may cease to be a significant factor in international relations after further development of nuclear energy, synthetic crude oil (syncrude), and synthetic gas (syngas). On the basis of available data, it is established that the Western world will cut its energy demand mostly through self-help projects and it will not depend unnecessarily on sources of supply beyond its control. A plan for energy independence or interdependence, based in the first instance on American national interest, could make a vast difference to international relations in the next 30 years and serve as a step toward an eternal world use of energy for the benefit of mankind.

Higher oil prices are forcing the energy consuming world, particularly the United States, to search for new sources of energy. Since lead time in energy development is long, planning for world energy independence can be comprehended in three overlapping decade-long time frames: (1) between 1975 and 1985 nuclear power will generate sufficient electricity by replacing oil and natural gas partially for other uses; (2) in 1985-95 syncrude and syngas developed from coal and other fossils will become an important energy; (3) during 1995-2005 solar energy will be utilized as an eternal source.

The central aspects of Saudi Arabian foreign policy as they relate to oil diplomacy vis-a-vis the Arab-Palestinian-Israeli conflict are elaborated in this study. There are four fundamental principles of Saudi Arabian foreign policy: (1) to prevent radicalist and nationalist incursion in the Arabian Peninsula; (2) to prevent communism from spreading its influence in the peninsula; (3) to maintain friendly relations with the industrialized non-Communist world; and (4) to support Islamic movement to promote solidarity among Muslim countries. The first of these objectives of the core of Saudi policy is designed to protect the Saudi monarchy against infiltration of radical nationalism into the kingdom. As an Arab nation it is impossible for Saudi Arabia to isolate itself from the tide of nationalistic and anti-Israeli sentiments in the region, which explains the reasons for the growing friendly relations of Saudi Arabia with Egypt, detente with Syria, and entente with the Palestine Liberation Organization (PLO). The second objective of Saudi foreign policy is to prevent the spread of communism in the peninsula. The atheism factor of communism is considered incompatible with Islam. The third principle of Saudi policy is to maintain friendship with the non-Communist industrialized nations. As a die-hard anti-Communist nation Saudi Arabia cannot offend for security reasons the United States, the most powerful anti-Communist country, and other non-Communist nations. Ideologically, Saudi Arabia has cast its lot with the anti-Communist world of its own volition, and the desert kingdom intends to remain as such. Being pronouncedly anti-Communist and underdeveloped, Saudi Arabia is dependent on the United States and other non-Communist nations for arms supplies and training, technology, and many other necessities of a developing nation, including vital supplies of foodstuffs. The fourth principle of Saudi foreign policy is to foster cohesiveness in the Muslim world. As the protector of the holy places of Islam, Saudi Arabia has assumed the role of a kind of spiritual leader of the Muslim world and, as such, wants to fulfill its resultant obligations to the Muslim nations. One such obligation is to liberate Jerusalem, the third holiest city of Islam, from Israeli occupation.

Saudi Arabia has remained an insular country, isolated from the main current of world politics, ever since the founding of the Saudi monarchy in 1932. From this utopian concept of insular living in the twentieth century, the Saudi kingdom has been exposed to the whole world after its decision to use oil as an instrument of foreign policy in the resolution of the Arab-Israeli conflict. The Arab policy to maximize political and economic gains through oil began in 1959 when an Arab Oil Congress was formed. Gradually, politics per se and politics of oil became inseparable and indistinguishable in the inter-Arab and Arab-Israeli conflicts. The Arab military incapability coupled with the magnitude of the American support of Israeli war efforts forced Saudi Arabia to use its oil weapon in the Arab-Israeli war. This use of oil as a weapon and its restrictions are examined from two opposite points: First, too little use of the oil instrument by the largest exporter and biggest "reservist" of oil in the world invites trouble for the Saudis in the Arab world; and second, too great use of the oil weapon provokes antagonism against the desert kingdom from the outside world, especially the United States. This dilemma, coupled with the prospect of development of alternative sources of energy and the growth of new economic and military power rivals, such as Iran, places Saudi Arabia in a very complicated process of petrodiplomacy which is elucidated in this study.

This study reveals a pattern in the control of oil in the Middle East as functional to the power relations of the day. Thus, with Britain relinquishing its role in the management of international relations early in this century, the United States assumed the responsibility, and hence the dominance of the Arab oil, only to find itself haggling with the rising power of the Soviet Union. In the meantime, in the formation of the Organization of Petroleum Exporting Countries (OPEC) in 1960, a power center, bent toward the correction of past injustices and control of the oil resources, appeared.

The cheapest and most abundant oil supplies are concentrated in the Middle East. Corollary to this oil concentration is the dependence of the industrialized nations on it. As the Arabs implemented their threat to reduce production and imposed an embargo on oil supplies to the world, they also stepped up efforts through diplomatic channels to achieve their objectives of getting an Israeli withdrawal from the occupied Arab lands. Since then the Middle Eastern states continue to be the scene of a vigorous struggle among the powers of the world.

The established American position has been challenged by the rising power of the Soviet Union and the growing independence of the Arab nations. While this rivalry was taking place between the superpowers, a new power center emerged: the Arab oil-producing countries had gathered together, forming the Organization of Arab Petroleum Exporting Countries (OAPEC) in 1967 for the purpose of reducing the power of the oil companies and controlling their own oil resources.

After the six-day war in 1967 Arab leaders urged the late King Faisal to use the oil weapon against the United States and other industrialized nations in order to force Israel to the negotiating table. So far Saudi Arabia's concern with Israel rested on pan-Arab considerations. Following the 1967 war, Israel expanded its boundary and Saudi Arabia no longer felt safe from Israeli threat. King Faisal, who had categorically ruled earlier that oil and politics should not be mixed, decided to use that same "unmixable" oil as an instrument of diplomacy to serve the Arab cause.

Because of the fear of Israel and for preventing the spread of radical incursion in the Arabian Peninsula Saudi Arabia emerged from its insularity by joining the militant Arabs to use oil as an instrument in the Arab-Israeli conflict. The Arabs perceived the role of oil as a weapon that could moderate the pro-Israeli policy of the major oil-consuming nations and isolate Israel from the main current of world politics. Again it was for this reason that Saudi Arabia shunned its policy of moderation. Saudi Arabia believed that, if Israel could be isolated physically and mentally from its neighbors and the rest of the world, the Arabs would have won and the Israelis would have lost a contest which otherwise has had no resolution through four wars. Saudi Arabia realized that it could accomplish this seemingly impossible task of isolating Israel only in association with other Arab oil producers. In this way Saudi Arabia could gain new power, prestige, and credibility in the Arab world, but not by remaining isolated from regional and global politics.

The enormous escalation of oil revenues gave the Saudis even more leverage with their oil weapon. For lack of absorptive capacity they simply do not need the additional cash flow that oil sales generate. The death of Nasser removed a towering personality from the Arab political scene and King Faisal could then act more independently, thereby retaining the control of the embargo in his hand as long as necessary. The oil war of 1973-74 was far more successful than any other attempt in the past in pressing the United States and other powers to come to the rescue of the Arabs in their effort to get back the occupied territories from Israel.

The Saudi motivation in using the oil as a political weapon was to kill not one or two but three birds with one stone: (1) to help speed up the process of Israeli withdrawal from the Arab occupied lands, particularly from Jerusalem; (2) to develop a diversified economic base of its own through cooperation with the Western industrialized nations; and (3) to assume a leadership role in the Middle East, nay, in the Muslim world.

The study focuses on the effectiveness of the Arab use of the oil instrument in accomplishing defined aims and discusses the extent to which it can expect future success. As it is, the possibility of another Arab oil embargo in the event of a renewed Arab-Israeli war cannot be ruled out. The impact of such an embargo on the Western and Japanese economies will be devastating. As such the United States has drawn up a contingency plan and made its intention clear that, if necessary, it would take over Arab oil fields to save the interests of the industrial consuming nations. The United States is not expected to remain passive if another oil embargo is imposed against it by the OAPEC nations. It is conceivable that the United States will respond positively through military, political, and economic actions to prevent strangulation of the industrial world. These and other ramifications of possible recurrent use of the oil weapon are illustrated in detail in the study. The options, limitations, and conditions surrounding the use of oil as an instrument of future Saudi policy are given due treatment in the analysis of the material.

Finally, two major conclusions are drawn from this study: Because of high probability for the development of the alternative sources of energy, oil will lose its present significance and will be reduced to the rank of a normal commodity in the world market. The age of oil is drawing to a close. Consequently, oil is losing its importance as an aspect of diplomacy. The other major conclusion is that oil may be used as a weapon of diplomacy in the short run in the partial settlement, but not the solution, of the Arab-Israeli conflict. In addition, attempts have been made to derive some useful general conclusions.

The methodology is a descriptive analysis with a combination of empirical data. The study draws on three main disciplines: politics, international relations, and economics. The emphasis of the study is on the political aspects of the oil embargo and the Arab-Israeli conflict. Using official documentations and statements of Arab-Israeli leaders and a substantial amount of literature on the subjects, including current periodicals and newspapers, the study is directed toward the determination of perception, motivation, and goal-definition of the parties to the dispute. Both primary and secondary sources are used in the study.

This is the first study on Saudi petrodiplomacy. Studies have been conducted on various aspects of oil diplomacy, mostly journalistic and topical, but no research work has been done on the use of oil as a weapon of diplomacy specifically as a case study of Saudi Arabia. Not only is this study original in context, but the analytical framework is also novel, which includes an analysis of Saudi "petrolism" within the Arab system, the OPEC system, and the world system at large. For convenience in interpretation, the world itself is divided into four worlds: The First World is the United States and other non-Communist industrialized countries, including Japan; the Second World comprises the Communist nations; the Third World covers the oil-exporting OPEC countries (although not specifically covered in this study, it is recognized that Third World covers many other fast developing nations); and the Fourth World includes oil-short and oilless preindustrial and nonindustrial countries.

The problem of spelling and transliteration of Arabic words and names is handled primarily by consideration of convenience, simplicity, and popularity. The generally accepted transliteration and spelling of

Arabic terms in the English literature is followed, except in cases of direct quotations and reproduction of documents. For example, Faisal is preferred to Faysal, Nasser to Nasir, Hussein to Hussayn, Jeddah to Jiddah, Koran to Qur'an, and ulema to ulama.

ACKNOWLEDGMENTS

This study originated as a dissertation, "The Use of Oil as a Weapon of Diplomacy: A Case Study of Saudi Arabia," at the School of International Service of the American University in Washington, D. C. It was completed in June 1975. Some changes have been made in the present study to update certain facts and make it more suitable for publication as a book. The author wishes to express his profound appreciation and gratitude to all those who made this study possible, and in particular:

To Dr. Alan R. Taylor, chairman of the dissertation committee, for his excellent guidance and assistance. To the members of the dissertation committee, Dr. Abdul A. Said and Dr. Darrell D. Randall, for their valuable suggestions. To Dr. M. Rashiduzzaman of Glassboro (N. J.) State College for suggesting the topic of research and offering excellent comments.

To Mr. M. Hossain Ali, ambassador of Bangladesh in Washington, for his repeated endeavors in postponing the author's transfer from Washington.

To his wife, Tutu, whose understanding, encouragement, and constant inspiration were invaluable in completing this study. And also to his sons, Bappi and Bobby, for their patience.

Finally, the author wishes to make it clear that the views expressed in the study are in pursuit of academic objectivity and do not reflect in any way his official position with the Embassy of Bangladesh in Washington. The dissertation committee chairman and members have guided the author in the preparation and improvement of this study, but they are not responsible for the views expressed herein. The author alone assumes full responsibility for his views.

CONTENTS

LIST OF TABLES AND FIGURES

LIST OF ABBREVIATIONS

Aramco	Arabian American Oil Company
IEA	International Energy Agency
LDC	Less developed country
MSA	Most severely affected (country)
NPC	National Petroleum Council
OAPEC	Organization of Arab Petroleum Exporting Countries
OECD	Organization for Economic Cooperation and Development
OPEC	Organization of Petroleum Exporting Countries
PDRY	People's Democratic Republic of Yemen
Petromin	Petroleum and Mineral Agency of Saudi Government
PLO	Palestine Liberation Organization
Syncrude	Synthetic crude oil
Syngas	Synthetic gas
Tapline	Trans-Arabian Pipeline
UAE	United Arab Emirates

LIST OF MAPS

Saudi Arabia and Oil Diplomacy

CHAPTER 1
SAUDI ARABIA

Saudi Arabia occupies most of the Arabian Peninsula which is slightly over a million square miles, or about one-third the size of the United States. The total area of Saudi Arabia proper is 865,000 square miles, with an estimated population of 8,115,000. There are no verified figures of the exact population as no accurate census has been taken.

Saudi Arabia is bordered on the north by Jordan, Iraq, and Kuwait; on the northeast by the Persian Gulf; on the east and southeast by Qatar, the United Arab Emirates, and Oman; on the south by People's Yemen and Yemen; and on the west by the Red Sea and the Gulf of Aqaba. The kingdom is strategically located between Africa and mainland Asia, lies close to the Suez Canal, and has frontiers on both the Red Sea and the Persian Gulf.

The area is notable for its desert climate. The main feature of the Arabian climate is the intense heat from April through October. Temperatures often exceed 120° Fahrenheit; the coastal area is usually hottest with high humidity. The moderate season from November through March is quite pleasant. A vast expanse in the southeast portion of the kingdom, known as the Empty Quarter, is too arid to support life. Saudi Arabia is one of the most bleak desert countries in the world. It has almost no rivers, no perennial streams, and no lakes. While 15 percent of its total area could be cultivated if water were available, in fact only about 1 percent of the land has enough moisture for agriculture of any kind. There are no forests at all. Rainfall is sufficient to support agriculture only in a limited area in the western part of the country.

The most fertile area is located in the province of Asir, in the southwest, adjacent to Yemen. The two principal sacred cities of Islam, Mecca and Medina, as well as the commercial center of the country, Jeddah, are located in the Hejaz. The province of Najd is situated in the heart of the kingdom; the eastern province of al-Hasa, bordering on the Persian Gulf, contains the principal oil installations.

Saudi Arabia is generally high on the west near the Red Sea, sloping down gradually toward the Persian Gulf on the east. The mountains that rise in places from the coast of the Red Sea catch some precipitation, but only in the southwest corner of the country is there adequate rain for agriculture.

PEOPLE

The people are basically Semitic of the eastern Mediterranean strain with African admixtures, particularly on the coasts, but all are identified as Arabs. The official language is Arabic, but English is taught in some schools in larger towns. Ninety-nine percent of the people are Muslims. The Saudi Arabs are predominantly Sunni Muslims who mostly adhere to the Wahhabism (puritanical interpretation of Islam). There are no native Christians, but many American, European, and Christians from other Arab countries are employed in the oil industry. Jews, except the current U.S. secretary of state, Henry Kissinger, and any other Jews in his entourage, are barred from Saudi Arabia, and non-Muslims are prohibited from entering the holy cities of Mecca and Medina.

Despite the concentration in Saudi Arabia of the world's largest oil reserve, it is still one of the most underdeveloped parts of the Middle East, and most of its inhabitants until recently lived at the subsistence level.[1] It is estimated that 50 percent of the population in the kingdom is nomadic.

With the establishment of the kingdom in 1932 and the coming of great wealth to the royal family and tribal sheikhs, a tendency toward class distinction has developed into an official line of class. The royal family, the House of Saud, is the highest class; then follow the lesser princely families, tribal sheikhs, and top ulema. A handful of wealthy members of successful merchant families has also attained upper-class life style.[2]

A new middle class is emerging rapidly in Saudi Arabia:

> This group of managers, administrators, technicians, clerks, teachers of modern subjects such as mathematics and science, lawyers, scientists, army officers and others in government and business occupy a middle level in prestige and socio-economic power, like other members of the middle class. But they are distinguished from the rest of the middle class by the reliance on secular, non-traditional knowledge to attain their positions. . . . The Saudi middle class also includes merchants, traders and landowners—the bourgeoisie—as well as middle level groups with a traditional education. Among the latter is an important group of Shariah judges, lawyers and religious scholars as well as teachers of religion and Arabic on all levels in the Saudi school system.[3]

The lower class is made up of nomadic bedouin, seminomadic herdsmen, unskilled and semiskilled workers in government and the private sector. Slavery legally existed in Saudi Arabia until 1963 when the government declared it unlawful and offered compensation averaging about $4,000 per slave.[4] Former slaves are at the very lowest level of society and are used extensively as infantry in the army.

The growth of the new middle class and its impact on Saudi life is hard to measure. The middle class is still numerically small. But the introduction of television, change to the Greenwich time system, adoption of an economic plan, use of the Gregorian calendar, spread of secular education, and public schooling of girls are innovations which show a dramatic growth of influence by the new middle class on Saudi life.

GOVERNMENT

Before the establishment of the Kingdom of Saudi Arabia, the peninsula had suffered for centuries from the struggle for political power, which rarely resulted in unity. The bedouin of the peninsula were always contending with one another over the control of the country, which had been devastated by their fighting. No one tribe in central Arabia had been able to prove its leadership, but tribes would alternately demand the support of other tribes when their leader was able to assert his authority over a larger area.[5]

During the early part of the eighteenth century, the preaching of a religious imam, Sheikh Muhammad Ibn Abd al-Wahhab, initiated a reformist movement in Arabia. Since then the central province of the Arabian Peninsula, the Najd, has been the home of the militant puritanical sect, the Wahhabis. Their ruler was expelled by the rival clan of Rashids from Najd to Kuwait. The exile's son, Abdul Aziz Ibn Saud, known in the West as Ibn Saud, reconquered Riyadh, the capital of Najd, in 1901. The present kingdom had its origin in 1902 when Ibn Saud returned to Riyadh from exile in Kuwait with a handful of followers. His followers, the fanatical Ikhwan who adhered to the Wahhabi sect of Islam, were eager to cleanse Arabia of what in their eyes were sacrilegious traditions which had been added to the religion of Islam.[6] In 1924 the Ikhwan entered Mecca and the next year took over Medina also. Ibn Saud was proclaimed king of Hejaz in 1926. In 1932 the various parts of Arabia were joined into the Kingdom of Saudi Arabia. The present kingdom has been independent since it was founded early in the twentieth century.

Since the establishment of the kingdom the political structure of Saudi Arabia has changed from a primitive tribal-religious patriarchy into an absolute monarchy in which the king's authority is limited only by Islamic law.[7] The government of Saudi Arabia is patriarchal rather than constitutional. Royal absolutism is tempered, however, because the monarch's power depends to a considerable degree on the support of the tribal chiefs.

The laws of Islam provide the country with civil and penal codes and also regulate religious problems. Religious law is the source of all legislation.[8] But the king has the power to issue a decree where the religious law is not applicable. The Koran is the supreme law in the kingdom. The Shariah, which is the foundation of the juridical system, is based on the Koran and the sunnah (custom, law, exemplary life) of the prophet Muhammad.

The judiciary is an independent agency in the state. The quadis (judges) usually judge according to the Hanbalite version of the Shariah law. Punishment for murder is decapitation by the sword. Justice, when not theoretically in conflict with the Shariah, can also be governed by tribal and customary laws. The late King Faisal observed:

> Our constitution is the Koran and our law is the Shariah of Mohammad (God's peace and blessing be upon him), our system of government is based on the interest of this country, where such interest does not conflict with the principles of our religion and the Shariah.[9]

In Saudi Arabia, the absolutism of the king has a great deal of benevolence. The king frequently holds open diwan (court), talking to his subjects, handing them gifts, listening to the requests and grievance of the people, and sharing with them the desert life of hunting and racing:

> The King performs such traditional tasks as holding open court on the palace steps and in the village mosques. These customs are extremely significant for the Sauds, for they are regarded by many tribal leaders as townsmen with tribal connections who attained their status only by conquests as Wahabi imams. The royal family must make special efforts to display the Bedouin Sheikh's attributes of courage, leadership, and generosity and must act, ostensibly, on the basis of tribal consensus.[10]

The government of Saudi Arabia is based on a tribal system and depends on the management of bedouin tribes by a display of authority understood in the Arabian desert. "Authority rests partly on respect for what is almost a religious interpretation, and partly on the ruler's own strict observance of Muslim law and sanctions."[11]

The king is the chief figure of the central government and the source of political power. He occupies the dual role of state and religious leader, manages the administration through the Royal Court, the Council of Ministers, regional and local emirs, and the tribal sheikhs. But the king is the final authority in all executive and legislative matters. However, on all levels government is administered on the basis of consultation; arbitrary decisions are rare.

Before 1932, King Abdul Aziz Ibn Saud directed everything himself, but later on the administration of the country went beyond the power of a single man. The country grew rapidly and new territories were added to the Saudi realm; it became difficult to exercise his authority without a flexible instrument of administration and permanent government:

> Although the government appears to be acquiring a western organization and structure, its operations in fact would hardly be recognized as such. Saudi Arabia is unprepared to staff this relatively complex organization. Higher posts are frequently taken over by the royal family in accordance with the tradition of trusting relatives above others. . . . Qualified Arabians are so few that the government is forced to hire Syrians, Lebanese, and Palestinians.[12]

Prince Saud took over the administration of the kingdom when his father (Ibn Saud) died in 1953. Gradually a quasi-constitutional system of administration began to evolve from a number of government decrees issued by the king on various matters, such as taxation, border control, and other affairs, within the framework of the Koran and the Shariah, which are still the basic law of the country.

A Council of Ministers was added to the traditional structure of Saudi Arabian central government. This body is a development of an earlier Council of Deputies established for Hejaz after its conquest by Ibn Saud. The Council of Ministers is responsible for the budget and other government affairs, including supervision of regional and local government, and is the supreme authority under the king. There has been a trend evolving from monarchical to ministerial rule. Regulations, treaties, international agreements, and concessions cannot be decreed by the king unless they have been prepared and approved by the Council of Ministers.

In 1958 during a financial crisis precipitated by irresponsible overspending, as well as by a political crisis with Egypt, King Saud was persuaded to permit his brother Prince Faisal, to take charge of governmental affairs as the prime minister. Although King Saud reasserted personal control in 1960, there was a continued struggle within the royal family from which Faisal emerged victorious in 1964. The royal family deposed Saud and proclaimed Faisal as the king.

Faisal's reign represented mild reforms, such as increasing the budgets for education and medical care. His reforms fell far short of those desired by the younger, educated generation. No political expression of any kind was permitted. The king worked on the assumption that the traditional system only needs to be made efficient and honest:

> He was determined to avoid an irreparable collision of Western materialism with his own Wahabite code, to lead his people to universal prosperity while insulating them from the mischief of

socialism, the godlessness of Communism and the decadence of the liberal democracies. He would bestow upon his kingdom all the blessings of technology, while retaining the Koran as the law of the land.[13]

On King Faisal's assassination on March 25, 1975, Crown Prince Khalid was enthroned as the king. The new king has vowed to follow the policies, both internal and external, pursued by King Faisal.

ECONOMY

Saudi Arabia is basically an agricultural and pastoral country, but owing to the scarcity of water, the small area of arable land has not produced enough for the people. There are no reliable figures available to determine how much of the land is under cultivation; however, agricultural production covers only a small part of the country, which is almost completely desert. The annual average rainfall is only five inches.

Apart from millet, wheat, and barley, the major agricultural product is dates, which are the staple of the nomadic people. With the exception of dates other agricultural products do not at all meet the total requirements of the country.

The present Kingdom of Saudi Arabia inherited a simple, tribal economy. Until the discovery of oil in the 1930s, Arabia's economy had been based on the oasis, the village, and the tribe—each small unit functioning almost independently. The discovery of oil changed the entire economic situation of Saudi Arabia. On May 29, 1933, the Standard Oil Company of California obtained a 60-year concession covering a large area in the eastern part of the country. An operating company known as the California Arabian Standard Oil Company was established. In 1934 when the Texas Company joined the enterprise, its name was changed to the Arabian American Oil Company (Aramco).[14] Although oil production started in Saudi Arabia in 1937 it was discovered in commercial quantities in 1938.[15]

One of the first steps of the oil concessionaire was to give a loan of £30,000 in gold sovereigns to the Saudi Arabian government.[16] Some authorities mention the amount of loan as being £35,000.[17]

The principal contribution of the first group of Aramco engineers was the construction in 1943 of a canal at the Kharj, an irrigation project near the capital, which carried pumped water a distance of ten miles.

During World War II, oil development virtually came to a standstill. The overseas movement of pilgrims, a most important source of foreign currency, fell off radically. To add to Saudi Arabia's difficulties, rainfall in the years immediately before the war was less than normal and the bedouins lost large numbers of sheep and camels.

By 1945 Saudi production of oil rose to 21,311 thousand barrels annually. The oil production registered an average growth of 6 million barrels a day in 1972. The total annual production thus rose to 2,201.8 million barrels in 1972.[18] The same year Saudi revenue increased by 45 percent and amounted to $2,734.1 million.[19] In 1973, the year of the Arab oil embargo, Saudi Arabian oil production reached 8.3 million barrels per day. Saudi Arabia accumulated $11.5 billion in reserves by September 1974.

The many years during which the income from oil was being squandered gave Saudi Arabia a later start than necessary in its economic development and planning. Officially, planning in Saudi Arabia started in 1958 with the establishment of the Economic Development Committee (EDC). On the recommendation of the International Bank for Reconstruction and Development a Supreme Planning Board (SPB) was formed in 1960 to replace the EDC. In 1965 a Central Planning Organization was established which replaced the SPB.

Under the impact of Western technology and through the cooperation of Aramco, Saudi Arabia is moving rapidly along the road to modern life. It is changing from a predominantly pastoral way of life to one depending largely upon agriculture and related industries. A few irrigation projects are already functioning and more are contemplated.

The discovery of oil has brought about transformation in the financial and economic situation of the country. The oil income has been put to some good use in establishing transportation and communication facilities, and developing water resources. The government has built schools and improved educational and public health facilities. Free facilities are available at all levels of education, and medical aid is also provided throughout the country free of charge, including treatment abroad for those cases which cannot be treated at home.

Until 1947 no budget estimates were published, but simple figures of expenditures and revenues were prepared for the purpose of negotiating with foreign powers. The first published Saudi budget was for fiscal year 1947/48, but only for general information. Since 1958, the Ministry of Finance has issued a yearly budget that differentiates the expenditure of the royal household and that of the state.

Serious development planning has been lacking in Saudi Arabia because of the dearth of reliable statistical information:

> . . . it can be said that economic planning of a somewhat imprecise and informal nature has been followed by the development-minded Kings in the postwar era. It has been planning by royal preference based in part in intuition, as well as on political and economic considerations, which has determined the allocation of the government's resources. A broad continuity has existed in the relative importance attached to the strategic outlays on current account, while the value attached to development projects has been greatly

> increased by the current regime. Furthermore, definite strides have been taken recently toward a more rigorous and formal approach to planning.[20]

In regard to capital investment, the most notable is the entry of the Saudi government into downstream petroleum refining and marketing operations through Petromin (petroleum and mineral agency).

The 1973-74 budget of £2,300 million was £611 million above the previous year, and 92.5 percent of the revenues came from oil.[21] In the decade following World War II Saudi Arabia has undergone many economic, technical, and social changes:

> While these economic, technological, and social changes are taking place, the political system of absolute monarchy based on Islam has remained unchanged. It is certain that sooner or later Saudi Arabia will have to face squarely the problem of adjusting its political structure to its socio-economic base. This adjustment may take an evolutionary or revolutionary road, depending on the foresight, skill, and statesmanship of the ruling class of the Kingdom.[22]

THE SAUDI ARABIAN OIL INDUSTRY

The scope of this section comprises political as well as associated economic aspects of the development of oil industry in Saudi Arabia. In this, the evolving relationship between the oil companies and the host country is surely most important.

The history of the oil concessions in Saudi Arabia dates back to 1923, when King Ibn Saud granted a concession, covering an area of more than 30,000 square miles in eastern Arabia, to the Eastern and General Syndicate, a British group; but since the syndicate lacked funds to cover the costs of exploration, the concession was allowed to lapse four years later. King Ibn Saud was in urgent need of funds. His principal source of revenue was the pilgrimage traffic, which declined from 200,000 persons in 1927 to 20,000 in 1933, as a consequence of the great worldwide depression.

The Saudi government's bargaining power was weak in 1933. After some rivalry with other interests, the Standard Oil Company of California obtained exclusive oil rights to a vast area in Saudi Arabia, which were subsequently assigned to the California Arabian Standard Oil Company. The original concession covered 360,000 square miles along the east coast of Saudi Arabia. The lease came into effect in July 1933 and was to run for 60 years; the terms called for a loan of £35,000 sterling and an annual rental of £5,000 sterling until such a time as oil was discovered in commercial quantities. The rest of Saudi Arabia remained unleased until 1939 when concessions were granted, which added 80,000 square miles to the original concession, thus bringing the total to 440,000 square miles and involving an additional cash payment of £140,000 sterling and a dead rent of £25,000 sterling per year.[23]

The success of the California Arabian Standard Oil Company (renamed Aramco) over its German, Italian, and Japanese competitors was received as a major victory in the United States. The United States, which had no diplomatic relations with Saudi Arabia, soon established relations with the kingdom. "President Roosevelt approved a proposal that the American Government purchase a controlling interest in the company; but the negotiations failed."[24]

Following the discovery and development of large petroleum reserves in Saudi Arabia, serious disputes arose after World War II concerning the financial terms of the concessions. As a result, Saudi Arabia was the first country in the Middle East to introduce the 50-50 profit-sharing formula. Once this principle was put into effect in 1950, it increased Saudi revenues immensely but cost Aramco little because the local tax could be credited against U.S. tax liability.

Several issues were raised in the following years in the application of the original concession agreement. Issues concerning the abolition of special discounts on posted prices, Saudi participation in the board of directors of Aramco, and the Tapline (Trans-Arabian Pipeline) revenue were resolved to the mutual satisfaction of both parties.

Apart from financial terms, another primary issue was that of relinquishment of large areas covered by the original concession. The original territory granted in 1933 to Aramco, together with the 1939 agreement to expand the concession, gave it a total of 496,000 square miles. By 1963, Aramco's concession area had been reduced to 125,000 square miles.[25] The remaining area was to be reduced gradually at five-year intervals, so that Aramco would be left with about 20,000 square miles in the last year of its operation.

The reduction in Aramco's exclusive concession enabled the Saudi government to negotiate new agreements with more favorable terms. Thus, oil concessions more favorable to Saudi Arabia were granted to Getty Oil Company (American), Arabian Oil Company (Japanese), AUXIRAP (French), and ENI (Italian).

As indicated previously, relations between Aramco and the Saudi government have been marked by such stress and strain, and it can be anticipated that there will be more in the future. In the meantime, Aramco has become a little more Arabian and less American, a process that began long before the 1973 Arab oil embargo.[26] It was noted in 1974:

> By the strict letter of its contract with the Saudi Government, Aramco owns concessionary rights to oil in the ground that is worth perhaps a trillion dollars at today's prices . . . , yet, submitting to Saudi pressure, the parent companies sold 25 percent of Aramco last year to the Government for some $500

> million. . . . This left 22.5 percent each for Socal, Texaco, and Exxon, and 7.5 percent for Mobil. Under last year's "participation" agreement, the Saudi share is to increase to 51 percent by 1982. But government officials have been talking of taking a bigger bite sooner, and not many oil executives would be surprised if Saudi Arabia eventually nationalized the company.[27]

It seems that, because of Saudi Arabia's physical control of oil, the only right Aramco and other companies have now is to produce, buy, and sell oil at terms dictated by the host government. However, relatively speaking, the relations between Saudi Arabia and the oil companies have been less acrimonious than in Iran or Iraq; although from time to time Saudi Arabia has brought strong pressure upon Aramco in order to modify the terms of the concessions.

The Saudis have complained about the little effective control their government wielded over Aramco. But following the application of the 50-50 principle, oil revenues to Saudi Arabia almost doubled within a year, rising from $56.7 million in 1950 to $110 million in 1951. Since then oil revenues have escalated continually, reaching a total of $2,734.1 million in 1972. This rise is due to higher oil exports and higher crude oil prices used in calculating Saudi income tax. Saudi Arabian oil revenues were estimated at $4.7 billion for 1973 and were expected to reach $21 billion in 1974.[28]

During the postwar period, foreign investment in oil has been the greatest in Saudi Arabia, and investment growth has been rapid. The operating company of Aramco started its activities in the kingdom in 1933, with an original capital of $500,000, which was raised to $700,000 in 1936. By 1946, Aramco had invested $115 million. In 1956, the total value of Aramco's gross fixed assets increased to $608.7 million. After the deduction of $269.4 million for depreciation and amortization, Aramco's net assets were $339.3 million by the end of 1956, which, after the addition of the company's net working capital of $67.7 million plus other assets of $64.3 million, stood at $471.3 million.[29] Since 1933 Aramco's total investment in Saudi oil production facility up to 1973 is $2 billion.[30]

There are widely differing estimates on investments in the oil industry in the world. But estimates on a country-by-country basis are not available except the figures collected here from various publications. This is because the oil companies provide the figures on their investments only on a regional basis. As part of their trade secret the oil companies do not divulge information on investments in individual countries. However, the U.S. Department of State compiles investment figures on a country-by-country basis. The figure of $2 billion cited above from a Department of State publication as the total of Aramco's investments in Saudi Arabia is most reasonable since the company has agreed to sell its establishments in the kingdom exactly at that price. Investments by other foreign companies are of recent origin and are considered negligible compared to Aramco's 42 years of most extensive operations in Saudi Arabia.

Given a discovered reservoir with a certain volume of recoverable oil under existing technology, any improvement in production technology increases the volume of recoverable oil at a certain cost. Volume increase as well as cost are by and large measurable and thus liable to rational calculations. Given the possible percentage spread between natural and stimulated recovery, there is a large incentive to perfect and to use improved production technology. This technology has had a dramatic impact on the oil production of Saudi Arabia.

The production of oil can logically be linked with exploration. In economic terms, the business of oil exploration, development, production, and transportation is fairly capital-intensive. This discussion includes a sketch of the specialized forms of long-distance transport used to move oil. Given a sufficient volume, pipelines are more efficient than other means of land transportation. Tankers and pipelines are used to move products as well as crude oil. But crude accounts for some three-quarters of the oil moved in international trade.[31]

Since our principal interest lies in the realm of political policy rather than in the economic and technical aspects of the Saudi Arabian oil production and transportation, the discussion of oil operations will be limited to a brief account of the experience of Aramco while operating in Saudi Arabia under a supplemental concessionary agreement, concluded six years after the signing of the original concession agreement with the Saudi government in 1939 for 60 years. Aramco is owned by four American corporations in the following proportions:[32]

	Percent
Standard Oil Company of California	30
Texas Company	30
Standard Oil Company (New Jersey)	30
Socony Mobil Oil Company	10

From the American point of view, Saudi Arabia is of primary importance, since it contains the largest "U.S.-owned" oil reserves and the concessions are entirely operated by American companies. Saudi Arabia ranks as the leading oil producer in the Middle East and the third ranking in the world after the United States and the USSR. Its proved reserves are the largest in the world,[33] about 132 billion barrels in 1974.[34] In 1972 Saudi Arabia produced 37.1 percent of the total oil production of all Arab countries and 22.6 percent of the total oil production of all members of the Organization of Petroleum Exporting Countries (OPEC).[35]

After oil is discovered in commercial reserves, several years are required before production, transportation, and marketing facilities can be set up. From a relatively modest beginning in 1933, Aramco's Saudi

Arabian venture developed into one of the greatest oil undertakings in the world. The first commercial oil field was discovered at Dammam in 1938; by the end of 1951 it was producing 90,000 barrels a day. In 1940 the Abu Hadriya and Abqaiq fields were explored and by the end of 1951 were producing 590,000 barrels a day. Qatif and Ain Dar developed in 1945 with a daily production of 20,000 and 150,000 barrels, respectively. Other oil fields were discovered in 1949 and 1951. By 1955, 154 wells in Saudi Arabia averaged 965,000 barrels of oil production per day.[36] Saudi production reached 8,290,000 barrels of oil per day before the renewal of the Arab-Israeli war in 1973.[37]

Aramco built a network of some 330 miles of pipeline, ranging from 10-30 inches in diameter and from 18 to 45 miles in length, in addition to the main Trans-Arabian Pipeline—the famous Tapline, 1,070 miles in length, to bring oil from the fields to the Bahrain refineries and to tanker loading stations. By the end of 1945 the refinery at Ras Tanura had been expanded to a 50,000-barrel daily capacity, and after further expansion the run to the refinery in 1954 amounted to 79,800,000 barrels. In general, Saudi Arabian oil exports are moved by pipelines to the sea terminal and delivered to tankers for transport to world markets.

After experiencing a slowdown in growth rates following the Arab-Israeli war of 1967, oil production has been registering a substantial increase as Table 1 indicates. The Kingdom of Saudi Arabia was instrumental for 60 percent of the total increase in free-world oil production in 1972, although the production of the Getty Oil Company, which was granted oil concessions in 1949 in Saudi Arabia, declined.[38]

TABLE 1

Saudi Oil Production, 1967-72

Year	Annual Production* (million barrels)	Daily Average (million barrels)	Percent Growth
1967	1,023.8	2.8	7.8
1968	1,114.1	3.0	8.8
1969	1,173.9	3.2	5.3
1970	1,386.3	3.8	18.1
1971	1,740.8	4.8	25.1
1972	2,201.8	6.0	26.4

*Includes the production of Aramco and the Saudi share of the production of both Getty and the Arabian oil companies.

Source: Compiled by the author.

The legal ownership of petroleum resources in Saudi Arabia belongs to the king. Petroleum operations in the kingdom have been developed and, until 1950, had been completely controlled by American oil companies under concessionary agreements reached with the king. The absence of general mining or petroleum laws in the kingdom necessitated negotiations and bargaining for the formulation of the terms and patterns of the concessions.

The concessionary arrangements provided extensive rights and privileges to the oil companies and guaranteed the continuation of their operations. Before the change that took place in 1950, a U. N. report commented on the agreements of the oil companies in the Middle East in general, which is equally applicable in the case of Saudi Arabia, as follows:

> The terms of their concessions . . . give the foreign companies a freedom of action which substantially insulates them from the economy of the Middle Eastern countries. Output is determined by considerations of world, rather than local conditions. Moreover, it is the companies which provide and own the means of transport, whether pipelines or tankers, to carry Middle Eastern oil to its markets and it is they who secure these markets, both in Western Europe and in other parts of the world. . . . The foreign exchange derived from sales of oil accrues to the petroleum companies and is in large measure retained by them. Hence the impact of oil operations on Middle Eastern producing countries is mainly indirect and the benefits consist of royalties and other payments of governments, foreign exchange spent by oil companies in the course of their operations, employment provided for workers, technical skills acquired by nationals employed by the oil companies and fuel provided for local consumption at reasonable rates.[39]

On December 30, 1950, Saudi Arabia and Aramco made a major change in the concession, under which Aramco was operating in the kingdom, by concluding a profit-sharing agreement, which is commonly known as the above mentioned 50-50 or "equal" profit-sharing agreement. "The agreement was of great importance inasmuch as it set a precedent for similar formulas in other oil-producing countries of the Middle East."[40]

By a royal decree on February 21, 1973, the participation agreement which was signed on December 20, 1972, was ratified. The agreement provided Saudi Arabia with an initial participation of 25 percent in the oil production, pipeline, storage, delivery, and export facilities. Under the agreement, Saudi Arabia is entitled to have an annual increment of 5 percent in its participation, starting January 1978, with a final increment of 6 percent culminating in an ultimate participation of 51 percent in 1982.[41] The agreement also gave the kingdom the right to play an active role in the management of Aramco. Aramco has moved the seat of its board of directors from New York to Dhahran. Saudi Arabia has so far made slight use of joint ventures, although it is heading in that direction.

Although the basic framework of the new concessionary agreements between the American, European, and Japanese oil companies and the Saudi government has remained unchanged, there have been a number of important changes in the terms of the original agreements in favor of Saudi Arabia, such as the one outlined above. These changes have been motivated by various political, social, and economic reasons aimed principally at obtaining a larger share of benefits derived from natural resources.

IMPACT OF OIL ON SAUDI ECONOMY

Before the discovery of oil, Saudi Arabia's economy had been very limited and was based on revenues from pilgrimage traffic, oases, and agriculture. Traditional characteristics of the Saudi economy were poverty, illiteracy, disease, and primitive techniques.[42] The outstanding source of revenue for the newly established kingdom in the early 1930s, as has been pointed out previously, was pilgrimage traffic which declined as the world was hit by the Great Depression, resulting in a worldwide economic crisis.

Saudi Arabia was rescued from this financial disaster by the American oil companies, which had already started their exploration. Since then "oil has become the dominant feature of the economy, providing some 85 percent of government revenue and 90 percent of the foreign exchange."[43] The petroleum industry has provided the vast majority of the Saudi national income and the highest wages. Oil royalties created changes and enabled the country to acquire many technological innovations. But there have been negative aspects:

> On the other hand, the sudden influx of wealth, concentrated in the coffers of the all-powerful Saudi family, also produced corruption, nepotism, and gross financial irresponsibility. Much of the oil income was wastefully dissipated. The royal family became a principal beneficiary. This large, extended group included not only the king and his children, but also spouses, children, and relatives by marriage, totaling hundreds, all of whom received direct grants. No longer were the Sauds austere Wahabi desert sheikhs. They now lived in oriental splendor surrounded by every conceivable luxury. Scores of royal princes purchased several high-priced American automobiles each and went on sprees of building palaces equipped with such modern amenities as air conditioning and swimming pools.[44]

The members of the royal family controlled vast investments in foreign countries, but little was done to better the lives of the people. As Prince Talal commented on the budget allocations in 1962:

> Of these budget appropriations, the greater portion of them was paid in the form of salaries to persons who performed no work at all such as members of the Royal Family and the heads of tribes. The King's allocations, for example, amounted to 166,000,000 riyals [one U.S. dollar in 1973 was equivalent to 3,7303 Saudi riyals] besides the salaries of his children which amounted to 18,000,000 riyals.[45]

A former oil minister of Saudi Arabia had this to say on the misappropriations of oil revenue before King Faisal took over the regime:

> Before the advent of oil, Arabia lived frugally, but everyone was content, now the poor man in Saudi Arabia is poorer. To add to this misery is the fact that oil revenue could naturally have made it easier for him to live decently, but fat profits reach the palace and never leave it. A small minority live in palaces equipped with the latest gimmicks, but the masses have no roof to protect them from the scorching sun.[46]

The majority of the people are poor peasants and workers, but a small number of elite rich families exist, and there is a small but growing middle class. The standard of living of the people has remained low, while that of the rich rose rapidly owing to the increase in oil revenues.

The many years during which the income from oil was being squandered gave Saudi Arabia a later start than necessary in its economic development. As noted previously, Faisal (who was proclaimed king in 1964) introduced reforms, such as increasing the budgets for education and medical care. In spite of the progress, Saudi Arabia still remains a country of striking contrasts. There are still the few wealthy and the masses of very poor people. The poor roam the desert in search of grazing land for their animals, getting little from the country's oil riches. In fact, the country is far behind most other Middle Eastern countries.

Oil income has made it possible to keep down the tax burden on the population and has also made possible a sharp increase in imports. There is a virtual collapse in the domestic market for local crafts, which imposes hardships on the countryside and throws rural areas into great dependence on supplies of imported goods from urban centers. As a result, the nomads are leaving the desert for the cities to seek jobs and shelter, which is causing some trouble for the Saudi government:

> Aside from the revenue it produces directly for the government, the oil industry affects the economy importantly in other ways.

MAP 1

Areas of Aramco Concessions 1933-1963

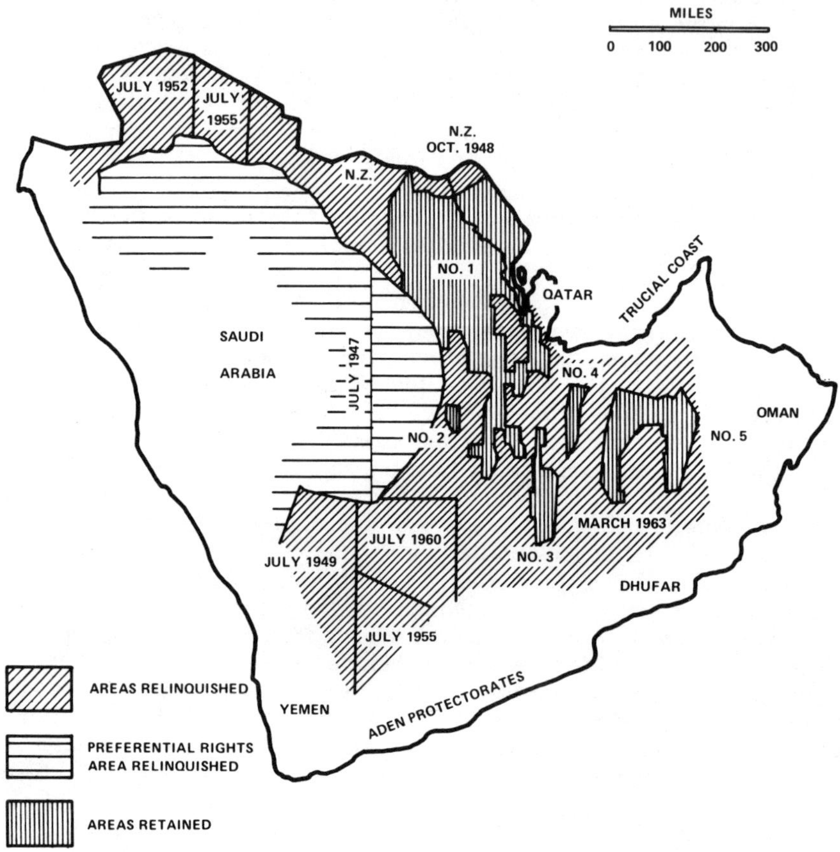

Source: Arabian American Oil Co. Aramco 1963, p. 12, reprinted from Ramon Knauerhase, The Saudi Arabian Economy (New York: Praeger, 1975) p. 160.

> . . . Since 1933 . . . the oil industry, Aramco in particular, has been the most striking feature of the economy. Operating mainly within the Eastern Province, Aramco . . . has established a huge, modern enterprise in the desert and has built whole towns and their ancillary services. The company has also transformed existing centers in the vicinity of its installations into prosperous and thriving communities, stimulated by the demands of company employees, but a lively import trade, and by the encouragement, technical advice, and support given to Saudi construction and contracting firms. Almost the whole life of the area revolves around the oil industry, which directly or indirectly determines the livelihood of the inhabitants.[47]

During the decade of Faisal's reign, the Saudi economy was carefully planned. The government assumed responsibility for the huge tasks of national education, creating infrastructure, marketing petroleum, erecting steel mills, acquiring a tanker fleet, and building plants to liquefy gas, produce petrochemicals, and assemble automobiles. The Saudi government is resolved to erect a vast industrial superstructure that will sustain future generations when eventually the oil resources dwindle.

There has been some progress, especially in the construction of public buildings and roads, but the problem of lifting a population that is predominantly illiterate and unskilled from dire poverty to a minimal standard of living is not an easy task. Even now, because of the inefficiency of the administration throughout the country and the continuing graft and corruption, much of the money budgeted for development is being wasted. "Today several billion Saudi dollars are sitting in the Chase Manhattan, Morgan Guaranty, and First National City Banks . . . "[48] Personally, "King Faisal's annual take has stood at about $50 million for years."[49]

Saudi society is far from egalitarian. Although the king has progressively reduced subsidies to the 2,000 princes, the royal family seems to be growing richer.[50] From the sale of oil Saudi Arabia was expected to accumulate $21 billion in 1974 and surpass the United States and Japan in monetary reserves.[51]

The Saudi government is seeking to conclude immense bilateral agreements with various Western nations and Japan to establish industrial complexes in exchange for steady supplies of oil. Simultaneously, Saudi Arabia has entered into an agreement with the United States to modernize and reequip its military forces with American training and armaments.[52] The Saudi government may also buy downstream oil refineries and service stations in the West, but it still lacks a coherent investment policy abroad.[53]

Although the great impact of the oil industry has begun to be felt in Saudi Arabia, its progress into the twentieth century is held back by a shortage of trained personnel. A weakness of Saudi Arabia has lain in the fact that it has never had sufficient trained men of its own to fill administrative and technical posts. Because of this lack, it has been compelled to depend on recruits from neighboring Arab countries and Pakistan:

> Sa'udi government would be impossible without two foreign factors, the oil revenues and the technical help of Egyptian and Palestinian advisers. Without the Egyptians and Palestinians there would be no civil service and no schools.[54]

As many Saudi Arabs receive higher education and training abroad and establish greater contacts with the outside world, dissatisfaction with life in the only absolute Arab monarchy grows. Many Saudis want more political freedom and greater guarantees of civil rights. Strikes and demonstrations are prohibited; only once, in 1953, 13,000 out of 15,000 Arab workers for Aramco in the eastern province went on strike demanding better living conditions and increased wages.[55]

The tide of Arab nationalism is rising in Saudi Arabia, and there is a nucleus of national opposition, which has been supported by Cairo as the Free Saudi Movement, with the following objectives, as stated by Prince Talal:

> The Saudi United Liberation Front formed with headquarters in Cairo, with an objective of the liberation of Saudi Arabia from its corrupt ruler and the establishment of a nationalist regime emanating from the people and affording social justice and equitable division of revenue. Its aim is to reach a sound, constitutional regime devoted to the benefit of the people and chosen by the people.[56]

But Faisal broke up the movement in the early 1960s by bringing its leader, Prince Talal, back home from self-imposed exile in Cairo.[57]

In the past, many Saudi military and air force officers have defected to Egypt, demanding close relationship with that country. But the most serious threat to Faisal came in 1969, when more than 200 air force officers and middle-level civilian officials were arrested after an attempted coup d'etat.[58]

Saudi Arabia is the world's largest producer of oil, after the United States in the free world; it is the world's largest exporter; its 132 billion barrels of proven reserves exceed any other nation's; its probable reserves may be much higher. One reason why Saudi Arabia, with its proven oil resources, has been unable to use its wealth effectively is due to the Wahhabi religious leaders. The ulema oppose Western innovations, such as cinemas and telephones. When the government opened the first girls' school in Buraida, the community did not want it, so they closed it. But as soon as Faisal heard about their action, he ordered the army to reopen the school. In 1963, a group of ulema and quadis delivered a solemn protest to the then prime minister, Faisal, when Radio Mecca employed its first woman announcer. They considered it a shameful act. Faisal replied, "I am as good a student of Islamic religion as anyone present," and asked, "So what is so shameful about women's voices now?"[59]

These are but a few examples of the friction and stress of change. Every day in Saudi Arabia there is constant dissent, such as between the father who disapproves of education and the son who likes to go to school, and between the bedouins who want to continue wandering in the deserts and the government which wants to settle them down.[60]

Development of industry is limited by the almost complete absence of domestic raw materials other than oil. Even then, with the establishment of light industries in the country, new problems have been created. The movement of a great number of workers into the cities has created problems, as housing and sanitation are inadequate. Moreover, the larger number of workers have increased the demand for labor unions, which are prohibited in Saudi Arabia, because the ruling family looks upon unions as a direct threat to the throne.[61]

The royal family is still strong internally. Despite the variety and persistence of nationalist attacks, revolutionary forces in the country are still weak, and the bedouins, who make up two-thirds of the population, and their leaders favor the royal family and the traditional way of Saudi Arabian life.

Trends toward secularization have been accelerated by the rise of new elements to positions of influence, but the process of secularization is energetically resisted by the conservative elements:

> The contrasts between the old and new in the Saudi economy suggest that the present period is a transitional one. At present, however, the most obvious impression is that the wealth from oil and the new economic activity which it has generated have been imposed on the old economic pattern without a satisfactory means of relating the two.[62]

Oil is by far the most important of the natural resources of Saudi Arabia, as has been discussed; the government does hope to spend $142 billion from 1975 to 1980 on welfare infrastructure, economic development, defense, and industrialization,[63] but in itself oil revenue is not enough to solve the economic and social ills of the kingdom.

The diffusion of the aspects of modernization—human, intellectual, political, economic, social, and psychological is neither so rapid nor so general.[64] A degree of clarity can be achieved only at the sacrifice of many essential interrelations. These are lacking in the oil-rich kingdom as has been pointed out. If nothing else, the waste of oil revenue by the royal family is a deterrent to the healthy development of society. Many Saudi Arabian observers believe that unless the king acts decisively and wisely to bring progress to the oil-rich backward desert kingdom, the country's next government will be a military or a republican

MAP 2

Other Oil Company Concession Areas

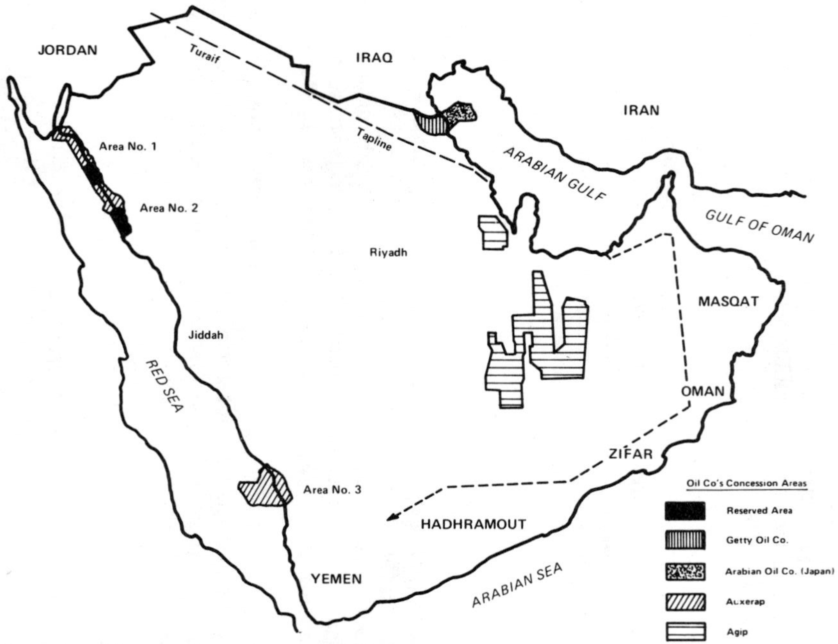

Source: Central Department of Statistics, Statistical Yearbook 1392 AH, p. 212, reprinted from Ramon Knauerhase, The Saudi Arabian Economy (New York: Praeger, 1975) p. 182.

one. Thus, it is for the opulence of oil that major political change may take place in Saudi Arabia. The obstacles to development in Saudi Arabia cannot be erased merely by increasing oil revenues to a point in excess of its capacity to spend the resources for modernization fruitfully.

It is difficult to evaluate precisely the effect of oil on the Saudi economy. The general impact of oil on the economy of Saudi Arabia has been considerable. In the absence of oil, some economists believe the situation would have been disastrous. However, statistical data suggest that development in non-oil-exporting countries of the Middle East did not score badly in the past compared to the oil-producing nations.[65] Without oil revenues, "it is probable that Saudi Arabia's economy today would not be appreciably different from what it was twenty-five years ago. . ."[66]

For the period since 1960, annual oil revenues have varied between 80 to 90 percent of total government revenues. The oil revenues have relieved Saudi Arabia of any appreciable foreign exchange restraint in endeavoring to promote development. But the Saudi capacity to utilize effectively its foreign exchange reserves has been less than its earnings. The abundance of easily earned money has created an atmosphere resulting in heavy dependence on imported luxury goods, an extravagant display of consumption, a sense of vanity, and little concern for efficiency in administration. A man of superior class never performs any manual labor in Saudi Arabia. Wealth is concentrated in the hands of a few, while inertia and an attitude of "let the government do it" overtake the majority.

Yet with the help of oil revenues, the government has erected seaports, airports, refineries, hospitals, clinics, and schools in the cities, as well as in the remote corners of the country. Education is the first priority and (as noted previously) it is free at all levels: " . . . a boy with any brains can count on a university education, with a salary while he studies."[67] Medical care is also free.

In the sociopolitical and cultural fields, a new concept of living has been born in Saudi Arabia since 1933. In this, Aramco has played a special and unique role. Oil constituted the wedge by which American culture, American attitudes and technology have intruded into the kingdom. "Perhaps nowhere in the world is the American method of free enterprise imitated with such devotion."[68] It has been noted also: "Given the earlier prevailing attitudes of Saudi Arabs toward foreigners, particularly non-Moslem foreigners, it is difficult to imagine Western penetration in the absence of the petroleum industry."[69]

The leverage afforded by oil revenue has profoundly affected the relationship between tribes and the House of Saud. In the past, one or another of the tribes have been in constant warfare and there was no central authority to maintain control over them. The oil income has helped the present dynasty to establish control over the tribes by paying the tribal leaders handsomely. This is all well and good, but its effects on output, when viewed as productive capital formation, remain to be seen. Petromin, the government-owned petroleum and mineral agency, has poured more than $200 million into industrial projects up to the beginning of 1974, but its records of achievements are not impressive.[70]

NOTES

1. Don Peretz, The Middle East Today (New York: Holt, Rinehart and Winston, 1965), p. 395.
2. William Rugh, "Emergence of a New Middle Class in Saudi Arabia," The Middle East Journal 27 (1973), p. 7.
3. Ibid.
4. Ray L. Cleveland, The Middle East and South Asia (Washington, D.C.: Stryker-Post Publications, 1973), p. 74.
5. W. R. Smalley, "The Wahhabis and Ibn Sau'ud," Muslim World 22 (July 1932), p. 227.
6. George Kirk, "Ibn Sa'ud Builds an Empire," Current History, December 1934, p. 291.
7. H. B. Sharabi, The Government and Politics of the Middle East in the Twentieth Century (New York: D. Van Nostrand Co., 1962), p. 225.
8. George Lenczowski, Middle East World-Center (New York: Political Institutions, Harper and Brothers, 1958), p. 123.
9. Embassy of Saudi Arabia, Prince Faisal Speaks (Washington, D. C., 1962), p. 41.
10. Peretz, op. cit., p. 402.
11. Olaf K. Caroe, Wells of Power (London: Macmillan & Co., 1951), p. 36.
12. Peretz, op. cit., p. 404.
13. New York Times Magazine, March 24, 1974, p. 14.
14. George Lenczowski, The Middle East in World Affairs (Ithaca, N.Y.: Cornell University Press, 1968), p. 548.
15. Roy Lebkicher et al., Aramco Handbook (The Netherlands: Arabian American Oil Co., 1960), p. 74.
16. Lenczowski, The Middle East In World Affairs, op. cit., p. 549.
17. Cited in Ibid.
18. Saudi Arabian Monetary Agency, Annual Report, 1972 (Jeddah: Research and Statistics Department, 1973), p. 3.
19. Ibid.
20. David G. Edens and William P. Snavely, "Planning for Economic Development in Saudi Arabia," The Middle East Journal 24 (1970), p. 29.
21. Encyclopaedia Britannica: Book of the Year, 1974. p. 606. The budget figures are shown in sterling as cited in Encyclopaedia Britannica. Saudi Arabian official figures are not available.

22. Lenczowski, op. cit., p. 572.

23. Sam H. Schurr and Paul T. Homan et al., *Middle Eastern Oil and the Western World: Prospects and Problems* (New York: American Elsevier Publishing Co., Inc., 1971), p. 116.

24. Ibid., p. 117.

25. Raymond F. Mikesell et al., *Foreign Investment in the Petroleum and Mineral Industries* (Baltimore: Johns Hopkins University Press, 1971), p. 225.

26. Allan T. Demaree, "Aramco Is A Lesson in the Management of Chaos," *Fortune*, February 1974, pp. 60-63.

27. Ibid., p. 63.

28. *Newsweek*, July 1, 1974.

29. Charles Issawi and Mohammed Yeganeh, *The Economics of Middle Eastern Oil* (New York: Frederick A. Praeger, 1962), p. 47.

30. Department of State, *Saudi Arabia: Background Notes* (Washington, D.C.: U.S. Government Printing Office, May 1973), p. 3.

31. J. E. Hartshorn, *Politics and World Oil Economics* (New York: Frederick A. Praeger, 1967), pp. 56-57.

32. George Lenczowski, *Oil and State in the Middle East* (Ithaca, N.Y.: Cornell University Press, 1960), p. 17.

33. George Lenczowski, ed., *United States Interests in the Middle East* (Washington, D.C.: American Enterprise Institute, 1968), p. 67.

34. *Time*, November 4, 1974.

35. Saudi Arabian Monetary Agency, op. cit., p. 14.

36. Arabian American Oil Company, *Middle East Oil Development*, March 1956, p. 33.

37. Leonard Mosley, *Power Play: Oil in the Middle East* (Baltimore: Penguin Books, 1974), p. 424.

38. Ibid.

39. United Nations, *Review of Economic Conditions in the Middle East*, E/1910 Add. 2/Rev. 1 (New York, 1951), p. 25.

40. George Lenczowski, *Oil and State in the Middle East*, op. cit., p. 19.

41. Saudi Arabian Monetary Agency, op. cit., p. 20.

42. Peretz, op. cit., p. 405.

43. Ibid., p. 406.

44. Ibid., pp. 406-407.

45. Prince Talal, "Saud Quo Vadis?" *Arab Observer*, August 27, 1962.

46. Abdullah El-Tariki, "The Makings of Deluge," *Arab Observer*, October 1, 1962.

47. George A. Lipsky, *Saudi Arabia: Its People, Its Society, Its Culture* (New Haven: HRAF [Human Relations Area Files] Press, 1969), p. 151.

48. *New York Times Magazine*, March 24, 1974, p. 58.

49. Allan T. Demaree, "Arab Wealth, As Seen Through Arab Eyes," *Fortune*, April 1974, p. 109.

50. New York Times Magazine, March 24, 1974, p. 54.

51. Demaree, "Arab Wealth, As Seen Through Arab Eyes," op. cit., p. 108.

52. Washington Post, April 6, 1974.

53. New York Times Magazine, March 24, 1974, p. 58.

54. Peter Partner, A Short Political Guide to the Arab World (New York: Frederick A. Praeger, 1960), pp. 94-95.

55. Sydney N. Fisher, The Middle East (New York: Alfred A. Knopf, 1959), p. 541.

56. Arab Observer October 1, 1962.

57. Washington Post, May 13, 1974.

58. Ibid.

59. New York Times, December 26, 1963.

60. Ibid.

61. Raphael Patai, Golden River to Golden Road (Philadelphia: University of Pennsylvania Press, 1962), pp. 338-45.

62. Lipsky, op. cit., pp. 154-55.

63. Washington Post, June 5, 1975.

64. C. E. Black, The Dynamics of Modernization (New York: Harper and Row, 1967), p. 56.

65. Charles Issawi, "Regional Development in the 1970s," in J. C. Hurewitz, ed., Soviet-American Rivalry in the Middle East (New York: Praeger, 1969), pp. 121-36.

66. Donald A. Wells, "Aramco: The Evolution of an Oil Concession," in Mikesell et al., op. cit., p. 230.

67. New York Times Magazine, March 24, 1974, p. 14.

68. Ibid.

69. Wells, op. cit., p. 234.

70. Demaree, "Arab Wealth, As Seen Through Arab Eyes," op. cit., p. 118.

CHAPTER 2
WORLDWIDE OIL SUPPLY, DEMAND, AND PRICE

PATTERNS, 1918-70

The expansion of the petroleum industry of the nineteenth century and of the early years of the twentieth century was not a war phenomenon. Rather it had a sound basis in peace and in the efforts of many peoples to raise their standard of living. The growing indispensability of oil has made its availability imperative. Mounting demand makes its production and distribution an attractive commercial proposition. Never has oil and its supply and demand been more seething than it is today.

The First World War transformed oil into a vital industrial and strategic raw material. For half a century petroleum had illuminated the world; now it was powering the machines of war. The tanks, the planes, the trucks, and even the taxies that helped the armies to fight the war were all moved by gasoline. Oil became a commodity of strategic importance as soon as various naval authorities realized oil's advantage over coal. "The war transformed oil from being a source of revenue for tycoons and speculators into a vital industrial and strategic material."[1]

When World War I broke out the two major producers were the United States and Russia. Certain quantities of oil were also being obtained from Indonesia, Romania, and Mexico. On the eve of the war oil had been discovered in commercial quantities in Venezuela. During the war Russian oil faded from the world picture.

In contrast to the United States with its ample reserves of oil, Great Britain had to contend with the absence of domestic oil reserves and a corresponding dependence on foreign sources. The British government gave diplomatic and military support to William Knox D'Arcy in his pioneering oil venture in Persia. The British government even became a part owner of the Anglo-Persian Oil Company. During this period the attention of the world oil industry was concentrated on the more obviously promising areas, such as the United States, the Caribbean, the

Middle East, and Indonesia. Europe and Africa were regarded with little interest.

As the war ended, the struggle for control of the Middle Eastern oil fields was followed by the formation of a cartel by the seven major oil companies. (The companies are Standard Oil Company of New Jersey, Standard Oil Company of California, Socony-Vacuum Oil Company, Gulf Oil Corporation, The Texas Company, Royal Dutch-Shell Oil Company, and British Petroleum Company. To these may be added an eighth, Compagnie Francaise des Petroles.) The efforts of the companies to establish a monopoly, or rather an oligopoly, over the production, refining, marketing, etc., of oil were aimed at establishing an international cartel to regulate prices and competition in such a way as to guarantee profits to all members of the cartel.

At the end of World War I, Royal Dutch-Shell was producing 84,000 barrels per day against Jersey Standard's 27,000. The two companies had a virtual monopoly in Egypt, Venezuela, and British Borneo; 96 percent of production in the Dutch East Indies; and 11 percent in Russia.[2]

Further: ". . . from 1922 onwards the United States produced more oil than its refineries could handle, and the refineries in turn pushed out more products than the market could absorb."[3] Important too were the new discoveries in Venezuela. From 2 million barrels in 1922, Venezuela's production rose to 9 million in 1924 and to 106 million in 1928, when it replaced Russia as the second largest producer in the world.[4] In 1930 vast oil fields were discovered in Texas, which dropped prices from $1.30 a barrel to $0.5 a barrel.[5]

The oil industry was established in several Middle Eastern countries before World War II. The total production of that region in 1938 amounted to about 15,000,000 tons. (See Table 2.)

TABLE 2

Main Oil-Producing Countries, 1938

Country	Million Tons
United States	161.9
Venezuela	27.7
Iran	10.2
Mexico	5.5
Iraq	4.4

Source: Christopher Tugendhat, Oil: The Biggest Business (New York: G. P. Putnam's Sons, 1968), p. 112.

As noted above, the First World War established petroleum as an essential war material. In World War II the need for petroleum and its products reached monumental proportions. In the railroad industry, steam locomotives had given way to diesel, and in the air the introduction of the DC-3 had led to the creation of a comprehensive airlines network. Without a doubt oil played a vital role in winning the war:

> In many ways the Second World War marks a turning point in the history of the industry, for after the war there was a steady decline in the control exercised by the major Companies, which was not apparent in the ten years or so following its end, but which became unmistakably evident in the later 1950s.[6]

Toward the end of the 1940s the United States became a net importer of oil. The result was the tying of low-cost Middle East oil to the high prices of U.S. oil, thus giving the major oil companies a magnitude of profit which allowed them to invest in large-scale downstream facilities from retained earnings. The major effect of the decolonization of the Afro-Asian countries and the creation of the United Nations, with its atmosphere of "one state, one vote," on the oil industry was a change in the rules of the game. This started off a round of discussions which culminated in the 50-50 profit-sharing formula. The first change in this system occurred in Venezuela in 1943:

> Originally, the rules and governments of these countries got some modest initial payments on granting a concession, usually combined with some rental for the territory covered by the concessions; thereafter, if oil was found in quantities which justified exploitation, amounts varying with the quantity of oil, i.e., a so-called royalty, became payable too.[7]

In 1950 Saudi Arabia became the first Middle Eastern nation to implement the 50-50 profit split, and by 1952 all other important producers in the region had followed suit, with the exception of Iran. Although the agreement between King Ibn Saud and Aramco gave the Saudi government half the profits attributed to crude oil production, it also limited the Saudi government's right to impose higher taxes, thus putting it in a less favorable position than the Venezuelan government.

Many of the Middle Eastern leaders have expressed their desire to alter the 50-50 profit-sharing principle to the advantage of their governments, a direct outcome of postwar changes in the market conditions of the oil industry. The survival of this method of splitting profits is not assured anymore since many factors which affected its evolvement have changed.

As a result of the large-scale relinquishing of acreage by the major companies, a proliferation of new types of agreements began to usher. With the exception of Soviet dealings with third parties, the new agreements are mainly between various national oil companies of the producing countries, which are viewed as symbols of the national aspiration to gain a place in the sun,[8] and either the established or new companies or the consuming governments. The newcomers include some American independents and various consuming-country companies. These new agreements have taken the form of equity participation, risk-taking, and vertical integration, among others:

> Both the major companies and the newcomers were successful at finding oil beyond their wildest dreams. In the late 1940s when the newcomers began to appear on the international scene, there were still lingering fears that one of the problems of the post-war world would be a shortage of oil, and reserves were sufficient to last for less than fifteen years at the prevailing rate of production. By 1960 there was enough to last for over forty years, despite the vast increase in consumption, and the industry's main problem was to find ways of preventing the flood of new discoveries from wreckage of price wars. [9]

TABLE 3

Oil Production by Main Areas, 1946-51
(million tons)

Year	U.S.	Caribbean	Middle East	USSR	World
1946	244	62	35	23	385
1947	262	69	42	27	424
1948	290	77	57	30	483
1949	267	76	70	34	482
1950	283	86	87	37	535
1951	322	97	96	42	608

Source: British Petroleum Co., Statistical Review of the World Oil Industry, (London: British Petroleum, 1956), p. 5.

World crude oil supplies were being boosted by major new supply sources (see Table 3). The great paradox of the international oil industry was that ever since the mid-1950s it had been facing a world surplus and declining prices. The reasons for this paradox were commercial and political.

The commercial reasons for the paradox have been described as follows:

> The industry is not a monolithic whole; it is a collection of different companies with conflicting interests which have to compete with each other in order to earn their profits. Some . . . have more oil than they need. But this is no consolation for Shell . . . nor to . . . the Italian ENI and French ERAP. For these groups the fact that there is a world surplus of oil is quite immaterial. It belongs to other people, and they want their own. Not even the reserve-rich companies

> can afford to rest on their laurels. . . . As . . . the key to profitability in the international oil industry is to have a source of supply as close as possible to every market.[10]

It is appropriate to underscore the comparatively recent nature of the political development. The political reasons for the paradox are based on past experiences with the Middle East, which is blessed with 75 percent of the world reserves and 46 percent of the production.[11] Six of the world's ten leading oil-producing countries in 1968 had virtually no recorded output in 1925. Many of the larger Middle Eastern oil discoveries date only from the late 1930s or later. It was only with the pervasive impact of oil supplies surging from the newer oil fields after World War II that established fuel patterns began their widespread change.

The expansion in output of oil in the Middle East and North Africa has been made possible by the discovery of huge oil reserves since the 1930s and by the continually rising demand for Middle Eastern oil in Europe, Japan, the United States, and other consuming countries. The large margin between receipts and outlays may be attributed to three factors: (1) The oil prices were sustained by the pricing system; (2) the various technical, institutional, and economic factors contributed to the exceedingly low costs of prices of oil; and (3) until 1950, when profit-sharing agreements were concluded, payments to the governments continued to be made mainly on the fixed royalty basis but although in most countries the absolute amount received by the governments increased, the rise in oil prices greatly reduced the producing government's share from oil income.

To comprehend further the problem of oil supply and demand on a global basis, the balance between production and consumption in various countries and regions must be considered. The countries which produce more than they consume must act as a source of supply for the countries which produce less than they consume.

In the 1960s some producing countries, notably the United States and the Soviet Union, were both exporters and importers. During this period when U.S. production exceeded domestic consumption, the Caribbean region was a major source of supply for Eastern Hemisphere requirements. Most of the Caribbean production came from Venezuela, which was second only to the United States in annual production. Although its production has been rising, the Western Hemisphere as a whole does not now produce enough oil to meet its own requirements.

Proven oil reserves are strikingly concentrated in a few major areas, especially in the Middle East. Therefore, the Western Hemisphere must look, increasingly, to the Middle East to make up deficits between demand and domestic supply. Japan is also dependent on the supply of oil from the Middle East. Indonesia contains some rich oil territory and is a substantial exporter. A series of major discoveries in North Africa and Nigeria indicates that these countries are destined to become important exporters of crude oil, especially to the nearby markets of Western Europe.

Oil prices had undergone a steady increase during World War II but following the end of the war prices declined sharply. During the 1950s the oil majors were successful in avoiding sharp changes in prices. With increased supplies of oil from North Africa and a rise in exports from the Soviet Union the companies reduced the posted prices of oil in 1959 by 18 cents and again in 1960 by 10 cents.

Two elements of interest in the pricing system are its rationale, and its implication. "Basically, it was a logical method for pricing oil in world markets under the supply conditions and industrial structure prevailing prior to the Second World War."[12] The system of posted prices, at which the companies stood ready to purchase crude at the wellhead, was tied to the income of the producing countries. Consequently, the companies maintained the posted crude prices but compensated their losses by juggling their books to show downstream losses in the consuming countries. The consuming countries resented both the ensuing loss of fiscal revenue and the fact that they were not compensated by a reduction of the foreign exchange cost of their oil imports. Acceding to pressure the companies reduced the posted prices for crude. While the reductions were not large their impact was a shock to the producing countries, and as a reaction they formed in 1960 their cartel—the Organization of Petroleum Exporting Countries (OPEC). OPEC was successful in preventing further reductions in the prices of oil which remained steady for one decade.

After years of collective bargaining between OPEC and the oil companies the system of posted prices was transformed into "tax reference" prices that, for all intent and purposes, could not be unilaterally reduced by the companies. A formula was also reached on royalties which gave a larger share of the profits to the producing countries. They were able to change the concession patterns by unilateral bargaining with the companies and not through the direct machinery of OPEC. An important aspect of the achievement is the large-scale relinquishing or confiscation of acreage held by the companies, giving the producing countries a source of additional revenue and also enhancing their bargaining power:

> It could be argued that all OPEC achieved had been on behalf of its Middle East members, not the others. Alternatively, you could rationalize it a little sourly by saying that anything which increased the tax-paid cost of Middle East oil helped Venezuela by making its oil to some extent more competitive.[13]

As a result of the Arab-Israeli war and the Arab oil embargo of 1967 and the closing of the Suez Canal, oil supplies from North Africa to the Western countries increased. The Western consuming nations became more dependent on supplies of oil from North Africa. Europe became more dependent on Libya as the civil war in Nigeria and the disruption of the Trans-Arabian Pipeline by Syria decreased oil supplies from Nigeria and Saudi Arabia. Libya alone was supplying about 30 percent of European requirements of oil.[14] (See Table 4.)

In 1970, Colonel Qaddafi, after taking over the government in Libya, cut back production of oil. As a result of the worldwide decrease in supply and increase in demand, the oil companies operating in Libya agreed to an increase of 30 cents per barrel in posted price of oil and 58 percent in taxes. These actions by the Libyan government and the oil companies triggered demands for new agreements on the part of other oil-exporting countries.[15]

The period between World War II and 1970 witnessed a steady erosion of the concessionary terms obtained by the oil companies. Until the beginning of 1970, the production and marketing of oil were largely controlled by the multinational oil companies. The situation began to change as many national oil companies were operating in the oil-exporting countries and as they became increasingly nationalistic. In this respect the main weapon of the producing countries is sovereignty and a near monopoly ownership. Without entering into a discussion of politically meaningless legal arguments this means that they have the power to act as they wish to control the production, distribution, and pricing of oil.

As a first step the development of oil supply, demand, and price in the world has been described above. Using this as a background, let us next examine the main features of oil supply, demand, and price in the 1970s. Particular attention will be placed on the role of OPEC and on the supply, demand, and price of oil in 1973—the year of the Ramadan or Yom Kippur war in the Middle East and the establishment of the Arab oil embargo which shook most of the world.

IN THE 1970s

Beginning with 1970, production was not increased to keep up with the ever-growing demand, and therefore, oil reserves in the United States began to deplete sharply. Although the demand for oil was growing in the United States, no important new discoveries had been made. In this, OPEC saw a good opportunity arising. Without wasting time, in December 1970 OPEC called for, among others, an increase in the income tax rate and in the price of oil. Alarmed by the decline in domestic oil production, the U.S. Department of State convened a meeting of the representatives of oil-importing countries in Paris for the purpose of convincing them that it was in their best interests to agree to the price increase.[16]

The circumstances were favorable for OPEC. At a meeting in Tehran on February 14, 1971, OPEC decided on a further increase in the price of oil. The other major features of the Tehran agreements, as described

TABLE 4

Oil Production in Selected Countries, 1963-70
(thousand barrels)

Year	U.S.	Venezuela	Saudi Arabia	Kuwait	Iran	Libya	Nigeria
1963	752, 723	1, 185, 520	594, 592	705, 471	538, 108	167, 788	27, 913
1964	2, 786, 822	1, 241, 657	628, 095	774, 815	618, 730	315, 660	43, 997
1965	2, 848, 514	1, 267, 709	739, 078	791, 903	688, 321	446, 493	99, 354
1966	3, 027, 763	1, 230, 464	873, 382	830, 537	771, 234	550, 185	152, 428
1967	3, 215, 742	1, 292, 873	948, 113	836, 719	947, 678	636, 504	116, 553
1968	3, 329, 042	1, 319, 340	1, 035, 773	886, 125	1, 042, 223	951, 427	51, 907
1969	3, 363, 832	1, 311, 827	1, 092, 322	940, 041	1, 231, 707	1, 135, 275	197, 224
1970	3, 517, 450	1, 353, 420	1, 295, 336	998, 110	1, 403, 558	1, 212, 049	395, 836

Source: Petroleum Press Service November 1973.

by Jahangir Amuzegar, the most noted Iranian authority on petrodiplomacy, were:

> First, the acceptance by the oil majors in February 1971 of a petroleum price policy based on "collective bargaining" instead of the previous practice of unilateral determination. Second, the introduction of an "escalator clause" or parity adjustment in income calculations—similar to the ones for wages based on general price level increases in industrial countries. Third, the establishment in December 1971 of a "real" price for oil based on putting oil on more or less the same parity with gold. And, fourth, the acceptance in March 1972 by the oil majors of the principle of local equity participation in oil-company assets and profits.[17]

In the Tehran Agreement of February 1971 and the Tripoli Agreement of April 1971, OPEC fixed prices for oil at the production stage and provided for a 2.5 percent annual increase in price plus five cents a barrel to take into account the rate of inflation.[18] These agreements were welcomed by the U.S. Department of State, which said that "the international oil business was entering an era of good feeling, one of stability that would last at least five years."[19]

The United States demonstrated its weakness as the largest consumer of oil by welcoming OPEC's decision to increase the price. Once this was done there was very little that could be accomplished to prevent subsequent price hikes. Iran, because of its credibility as a strong pro-Western nation in American eyes and as an uninterrupted supplier of oil to the Western nations and Israel, could take the initiative in the Tehran and Tripoli agreements. "It was Iran that took the lead in asking for higher tax rates immediately after the early Libyan agreements, and Iran was the first to get a price increase for heavy crude oils."[20]

After a slow growth in 1971, crude oil production received a new impetus in 1972 when production in all countries reached 2,598 million metric tons, about 52 million barrels per day, or about 5 percent more than during the year before. The year's growth was concentrated in a relatively small number of countries, foremost among which were Saudi Arabia and Iran.[21]

On the one hand Iran increased its oil production to take advantage of the higher price by selling more, and on the other it took the lead again in the Geneva Agreement in January 1972 to raise the price of oil by 20 cents per barrel to offset U.S. dollar devaluation in August 1971. Since then, OPEC at a meeting in June 1973 raised the price of oil by another 15 cents per barrel, again to offset U.S. dollar devaluation in February 1973.[22] In August, Iran wiped out the 19-year-old consortium agreement covering onshore oil production. What was forthcoming was still another oil price hike. In September, OPEC decided to introduce a considerable increase in posted prices, and thence in royalties, effective as of October 16. The OPEC nations also "called for a renegotiation of the Tehran Agreement, noting that the current rate of inflation far exceeded the 2.5 percent rate contemplated in that Agreement."[23] Both OPEC and the oil companies agreed to negotiate, but the negotiation was never completed. In October OPEC members in the Persian Gulf, led by Iran, unilaterally increased the price of oil, raising it from the then-existing price of $3.01 per barrel to $5.11. "For the first time, oil-exporting nations took the pricing function completely into their own hands."[24] Thus the established process of negotiation between OPEC and the oil companies ended abruptly. It was an overt political act on the part of OPEC, not an economic matter, to end prematurely the process of negotiation unilaterally in violation of their own agreement reached in Tehran in 1971

A fundamental change in the structure of the oil industry took place following the OPEC decision to do away with the principles of the Tehran Agreement which were supposed to last for five years, that is, until 1976. Now the producing countries could dictate the terms of pricing and other conditions of oil sales at will. In this Iran has been in the forefront, rather at the expense of the Arab nations. The Iranian chairman of the gulf "chapter" of OPEC was present at the historic meeting of the Organization of Arab Petroleum Exporting Countries (OAPEC) in Kuwait on October 17, 1973.[25] The very fact that Iran, a non-Arab nation, was present at the meeting of OAPEC, an exclusive Arab oil cartel, was indicative that it had some motives. As subsequent events showed, Iran indeed had such motives.

At the beginning of the OAPEC meeting in Kuwait, Saudi Arabia successfully resisted attempts to set an embargo against the Western nations. Even Kuwait announced that it would contribute $350 million to the warring Arab states' efforts but that it did not want to stop oil output.[26] It seems that the Iranian motive to attend the OAPEC meeting was to influence Arab nations to cut off oil supplies to pro-Israeli nations. The decision to cut off the production of oil was announced by the Iranian chairman of Gulf OPEC before the handwritten Arabic text of the resolutions of Arab oil ministers was distributed to newsmen.[27] Maintaining its posture of friendship toward the Western nations and at the expense of the Arab nations Iran wanted to increase oil production and sell it at an exorbitant price to the industrialized world and even to Israel. Taking advantage of the embargo, Iran proposed to increase the posted price of oil to $23 a barrel at a meeting of OPEC in Tehran in December 1973.[28] At the instruction of King Faisal, Saudi Arabian Oil Minister Sheikh Yamani not only resisted the move but clashed with the Shah of Iran, who finally settled for a price of $11.65 a barrel, effective as of January 1, 1974.[29] This was the officially established price, but as happens in crisis situations, Iran sold oil at more than $16 a barrel at auction while the embargo was in effect.[30]

The energy shortage became a problem of major proportion in many parts of the world during 1973. Especially hard hit were the industrialized nations of the West and Japan. The problem concerning fuel supply was caused basically by the continuing rapid growth in demand for energy, particularly in the form of petroleum products. For example, oil consumption in Western Europe rose sevenfold from 1956 to 1973, when the Arab countries of the Middle East and North Africa cut back production in retaliation for the Western, principally U.S., support of Israel during the Arab-Israeli war. The Arab countries virtually declared an "oil war" on the West and Japan. The oil war brought in its wake a whole chain of diplomatic activities which will be discussed in Chapter 6.

The seriousness of the energy supply shortage was underscored by the fact that in 1973 Western Europe depended on the Middle East for more than 70 percent of its oil and Japan relied on the region for over 80 percent of its supply. The United States was much less dependent on imported oil, but the Arab oil embargo, nonetheless, had a considerable impact on America.

Many critics in the United States believed that the energy crisis had been contrived deliberately by the large oil companies in the interest of greater profits. Although the companies denied this charge, it appears that, in part, the energy crisis is a reflection of political and economic pressures exerted on the oil industry as early as the 1950s. After the Suez crisis the Iranian parliament passed its Petroleum Act of 1957, which for the first time recognized and encouraged the concept of joint ventures between foreign producers and the host country in the Middle East.[31]

When other nations followed Iran's lead, the oil companies realized that they would not only be called on to pay more money but gradually to cede control of oil operations to the producing countries. As a result of this change, the oil companies began to diversify their holdings in search of oil in other regions of the world, such as Southeast Asia and Australia. They also moved in Alaska and Canada and increased their drilling on the outer continental shelf.

Soon the Middle Eastern and Caribbean countries became increasingly militant, adopting nationalistic policies, which resulted in the assumption of greater ownership of the concessions being operated by the foreign oil companies. Additionally, the host countries began to increase the price of their output. The oil companies, whether in protective reaction against possible nationalization or through a natural rapacity for acquisition, began to shift their strategies. Most important of all, since the beginning of the 1960s, the oil companies began to acquire and control the development of the competing energy sources, such as coal, uranium, nuclear energy, and natural gas. The major American oil companies also acquired oil shale and tar sands, as well as water rights in many parts of the United States.

If the controversy over the development of the competing sources of energy was at the base of the energy crisis, for most people the crisis was translated into higher prices and a threat of scarcity of gasoline in the autumn of 1973 and fuel oil in the winter of 1973-74.

The export prices of oil have roughly quadrupled since October 1973. Table 5 shows the components of the effective price of crude oil on world markets for three dates in 1973-74.

TABLE 5

Price of Crude Oil, October 1973-January 1974
(dollars)

Price Component	October 1, 1973	October 16, 1973	January 1, 1974
Posted price	3.01	5.12	11.65
Royalty	0.38	0.64	1.46
Production cost	0.12	0.12	0.12
Income	1.38	2.40	5.54
Government revenue	1.76	3.04	7.00

Source: *Science* 184, no. 4134 (April 19, 1974), p. 322.

Based upon 1973 production and prices, the economic rent collected by the OPEC countries amounted to about $25 billion per year, growing at 11 percent or more yearly:[32]

> Even before the dramatic increase in oil prices in October 1973, the Middle Eastern oil industry played a very important, but quite discreet, role in international money flow. By the mid-1960's the presence of U.S. owned oil companies in the Middle East and the marked preferences of most oil exporting countries for U.S. produced goods and services combined to generate large net flows of dollars into the United States.[33]

High oil prices have a relatively minor impact upon industrialized countries, since they can borrow to finance their oil imports until broader adjustments ensue. The Fourth World nations have no such options and their economic hardship is not considered fundamental to the global economic order. As MIT economist Morris A. Adelman points out, "it's the underdeveloped countries that certainly are hurt as much or more than anyone by high oil prices, but they don't say a word."[34] Sooner or later the Fourth Worlders are likely to speak out to seek redress for their grievances, if the hopeless situation continues for long. Realizing this, the oil exporters have begun to soften their attitude toward the Fourth World and are giving economic assistance, in some cases in greater volume and more per capita than the Western nations had done individually in the past.

Perhaps the hardest hit developed nation was Great Britain. There, bans on overtime work by coal miners, electric power workers, and railway engineers were added to the oil cutback to create a condition of crisis and national emergency. Other Western European nations and Japan were also compelled to enact fuel-saving measures. Conservation efforts began taking a toll on various sectors of the U.S. economy.

The escalation in oil prices has had a dramatic effect on international trade balances and monetary debt. The higher oil prices have caused trade deficits in almost all oil-importing countries. While the rich nations with economic flexibility have been able to adjust to the shock of oil price revolution the poorer countries in the Fourth World are accumulating their huge debt burden. By one estimate the total international debt caused by the oil price increase is $55 billion. Of this amount, the share of the poorer countries is as much as 40 percent.[35] To cope with this situation the International Monetary Fund (IMF) has established a special oil-lending or recycling facility with funds provided by OPEC. So far OPEC has contributed $9.6 billion for this purpose. Nearly all of this amount is being lent to Fourth World countries. The developed countries have created within IMF a $25 billion full "safety net" fund for the 24 nations belonging to the Organization for Economic Cooperation and Development.

The oil price increases in the early 1970s were due to natural reaction to an inflationary world economy and devaluation of the U.S. dollar. But the price hikes in late 1973 were motivated by political consideration for economic gains. In this Iran played a decisive role. First, by unilateral rejection of the process of negotiation with oil companies, a situation was created in which the oil exporters could increase the price at will. Second, by help to the Arabs to establish the oil embargo, an artificial shortage was created in the world market which was further conducive to price escalation. Finally, to sustain the increased price level, Iran has consistently exercised its influence against the efforts of Saudi Arabia to bring oil prices down. The Iranian objective is to maximize revenue, which involves emphasis on price, while Arab priorities relate to the resolution of their conflict with Israel. On this Iranian-Arab conflict of interest George Lenczowski has the following to say:

> The spectacular quadrupling of oil prices was initiated by Iran, a non-Arab country, at a meeting in Teheran in December 1973 and was adopted by other member-states of the Organization of Petroleum Exporting Countries (OPEC) for reasons that only had nothing to do—or little to do—with attainment of Arab boycott objectives, but actually contradicted it. Although the Arab production cutback policy was aimed at penalizing foes and rewarding friends, the drastic price hike hurt friend and foe alike, thus nullifying to a certain extent the benefits expected to be gained from the boycott strategy.[36]

At this writing, in April 1975, as a result of the recession in the United States and the conservation measures practiced by all oil-consuming nations, worldwide production of oil has exceeded demand by 4 million barrels a day (see Figures 1, 2, and 3). To sustain the higher prices, oil-exporting nations are trimming their production. As inventories of oil are being built up according to schedule in the industrialized countries and as practical substitution programs begin to take shape, allowing oil to remain in the ground for an extended period of time is no longer a profitable proposition. This, coupled with the psychological insularity of OPEC nations, has already produced some desired effects on the price of oil which are showing some signs of decrease. But it is unlikely that the price of oil can be reduced. Rather, it is likely to increase in the near future, since supply-demand ratio does not determine the price of oil anymore; it is decided by the wish of the OPEC nations.

Before the Arab oil embargo, cost and supply from the Middle East "determined world oil prices as well as prices of other energy sources. In the future, U.S. demand and other energy costs will set the pace in world oil prices."[37]

Politics aside, it is because of the approximately one-third consumption of the total world energy requirements by the United States that the oil-producing nations are encouraged to escalate the price of oil, knowing that the United States can afford to pay the higher price. By extracting the American dollar through the sale of a product which can eventually be substituted and which could already have been accomplished if timely U.S. action had been taken, oil-producing nations are in a manner challenging the affluence of the United States—affluence developed mostly through technological innovation and self-help. When the day arrives that the United States is able to stop oil importation, the price of oil would fall, due to the lessening of demand.

COMPANY-HOST-CONSUMER RELATIONSHIPS

Historically, oil and politics are an old mixture. A brief recapitulation here of the oil concessionary system in Saudi Arabia and company-host-consumer relationships may shed further light on oil and politics. Saudi Arabia granted oil concessions to Aramco in times of extreme weakness and poverty, allowing generous privileges and maintaining practically no control over company operations and policies. Negotiations were conducted "in camera" and were not made public; however, a change began as the surge of economic nationalism was engulfing the Mideast.

The apparent reluctance of the oil companies to cooperate with Saudi Arabia in the past created an atmosphere of mutual suspicion. The situation began to change as Saudi Arabia started to look for diversified

FIGURE 1

1973 Crude Petroleum Production and Petroleum Product Consumption for Major Producing and Consuming Areas (million barrels per day)

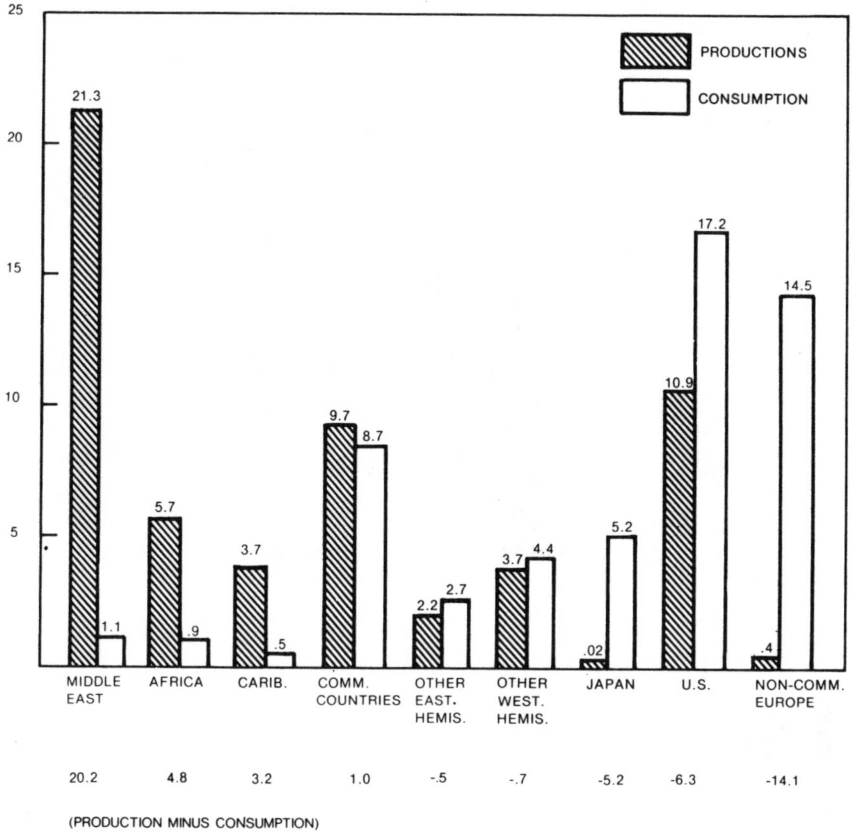

Source: Federal Energy Administration, Project Independence Report, November 1974.

FIGURE 2

1973 Import Levels and Proportion of 1973 Imports from Middle East for Major Importers [MB/D]

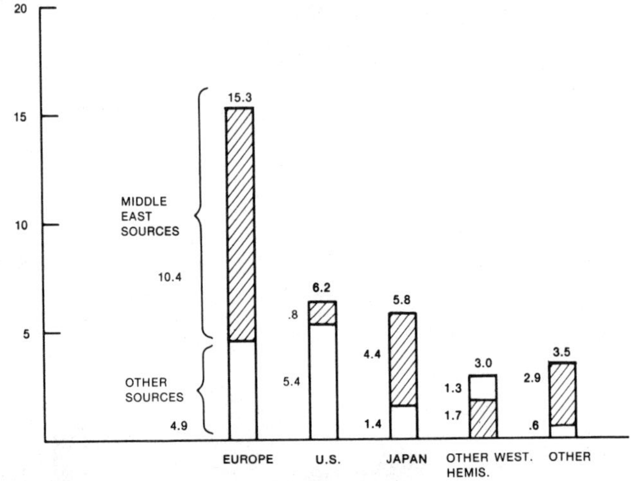

FIGURE 3

1973 Crude Petroleum Reserves for Major Producing Areas

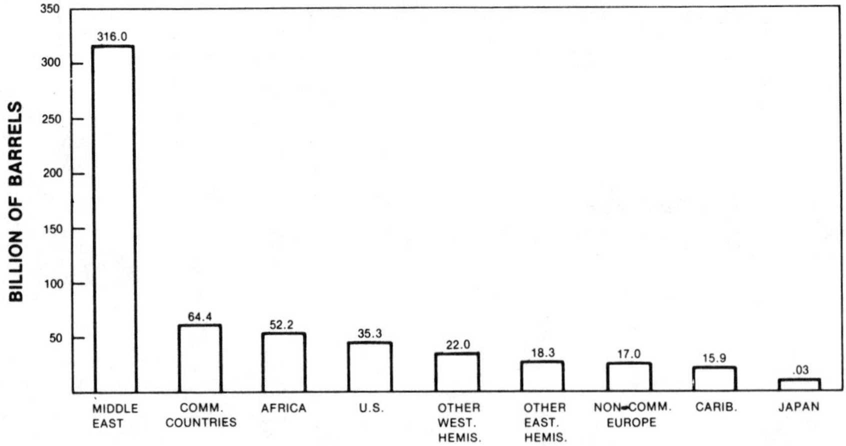

sources for oil concessions. Slowly the French, Japanese, British, and Italian companies were entrusted with oil deals. At the same time Saudi Arabia commenced to patronize its own national oil company.[38]

Sensing this gradual change in the Saudi attitude toward oil companies Aramco adopted a posture of becoming a little more Arabian and helped the Arabs in the implementation of their embargo against its home government. In addition, Aramco even backed the producers' policy of the oil price hike.[39]

In the past the oil companies dictated the terms of agreements to the host governments. In other words, the consuming nations guided the policies of the oil-producing countries. Now that the producers are united under the banners of regional oil organizations they are determined to maintain the control of their economy at their own hands and cooperate with outsiders only on terms dictated by their own self-interests.

Basically, this changing pattern of company-host-consumer relationships involves three actors: the exporting countries, national and multinational oil companies, and the consuming governments. During the early 1970s the demand for oil increased in the major consuming countries as Table 6 indicates.

TABLE 6

Oil Imports to Consuming Countries, 1970-73
(thousand barrels daily)

Country	1971	1972	1973	Rise: 1973 over 1968 (percent)
United States	3,930	4,740	6,205	17.2
Western Europe	13,520	14,065	15,310	7.9
Japan	4,720	4,815	5,760	13.5

Source: British Petroleum Co., Statistical Review of the World Oil Industry, London, 1973.

The oil companies met this increased demand from existing sources of supply. They were reluctant to explore and exploit new reserves of oil as the producing countries were increasingly opting for participation or nationalization of their oil resources. Originally Saudi Arabia devised a 25 percent participation arrangement, which was soon followed by other gulf states. The share of participation was increased from 25 to 60 percent and talks are underway about the terms on which the Saudis will take over the remaining 40 percent of Aramco's ownership.[40] The full participation agreement will be, in effect, a bid to make Aramco a mere production contractor. Apart from other immediate effects, this trend will represent a broad movement among oil-producing Third World countries to arrange for a full participation agreement or complete nationalization. The assumption here is that the Saudis will achieve 100 percent ownership and other oil producers will follow that pattern, or vice versa.

Taking further advantage of the changed political realities in the world, the oil-producing Third World, in association with the Fourth World, which was joined by an overwhelming majority of the U.N. members, adopted a Charter of Economic Rights and Duties of States, proposed by the president of Mexico. The General Assembly adopting the economic charter by a vote of 120 to 6 proclaimed on December 13, 1974, the right of poor countries to control their own natural resources and take over foreign firms.[41] The United States voted against the measure. Not only the days of American dominance of the United Nations were gone but also American control and influence in the world oil industry. On the same day the OPEC nations, meeting in Vienna, decided to replace the multiple pricing system of oil by a uniform flat price. The key point of the new system is to eliminate the posted price which was only a reference figure to compute taxes and royalties. At the same time the price of oil was increased by 38 cents per barrel, which increased the producing governments' revenues from $9.74 a barrel to $10.12.[42]

The real effect of the new system will be to shave the profits derived by the multinational oil companies. In the past, the companies have compensated for such a cut in their production profit margins by passing the buck to the consumer in the form of higher prices:

> In theory a high price suits a commercial firm engaged in expensive operations all over the world provided that its profit margin remains; but though in fact the companies' net incomes rose enormously thanks to the 1974 prices, they were worried about the reduced demand these prices presaged, and about the changing shape of their industry.[43]

In any event the oil companies had no alternative but to concur with the price hike and maintain a low profile in the host countries. These are the days of great uncertainty for the oil companies and, in their attempt to survive, they are diversifying their operations into substitute energy and other fields. A few examples will suffice to show the adaptability of the companies. Shell and Exxon have entered into both the nuclear and the coal business, while Mobil has bought the American mail-order business of Montgomery Ward.[44] Atlantic Richfield Company (ARCO) is conducting a huge research project on the need of public transportation with the hope of developing business in the transport sector. Oil companies are even diversifying their operations abroad. Aramco has undertaken an electrification program for the whole of eastern Saudi Arabia. Aramco is also designing, building, and operating a $4.5 billion gas and gas-liquid system to serve planned industrial centers in Saudi Arabia.[45]

The problem is, how can the nations of the world handle the new situation? To survive, the oil companies are making alternative arrangements. The amount of the oil producers' income is almost anybody's guess. The increase of 38 cents a barrel which went into effect on January 1, 1975, will add another $1.5 billion annually to the oil producers' coffers.[46]

Even at the prevailing prices before this new increase, "At minimum OPEC as a whole, after satisfying its import needs, disposes of a surplus of . . . $650 billion by 1980 and $1,200 billion by 1985."[47]

Worse hit than the industrial consumers are the 32 so-called most severely affected (MSA) countries, including the nations of South Asia and African Sahel. Though their per capita consumption of oil is small, most have such large populations that to satisfy their oil needs their financial resources are being depleted so fast that bankruptcy seems possible. Bangladesh is, perhaps, the most notable example of this stagnating group of nations. At the other extreme is the United States, which, geographically speaking, is practically a world in itself. It could probably remain a host in itself, requiring little outside help, but, because of its unwise consumption and waste of energy coupled with the lack of adequate initiative to develop substitute energy in time, is dependent on external oil supplies. Bangladesh, one of the newest and poorest nations, is becoming insolvent paying for its food and petrobills, while the United States, the oldest of the newest and one of the financially strongest nations, is becoming ill. The American illness can be cured, of course, by utilizing nuclear, coal, and solar energies.

In sum, the attempt by the oil producers to cripple the small and the giant is a serious challenge to the world's oil supply and financial systems. The oil companies, principal actors in the past, are silent spectators now. The attempt by the oil-producing countries to link the price of oil with the prices of other major items of trade, such as foodstuffs and manufactured goods, will make the poor nations even poorer. Although the alleged intention of the oil producers is to protect their oil revenues against inflation in the industrial countries, their attempt to index oil prices with other essential commodities without regard for the cost of production and market conditions is striking a severe blow to the poor countries' survival. Like dying patients in a hospital, unable to talk but staring at visitors and medical attendants with the hope that medical relief will be forthcoming, the poorer countries are silently looking at the oil-producing nations, hoping that relief will come at the eleventh hour as a salvation from collapse. The new relationships between the producing and consuming governments have become those of confrontation. As a result, the world is facing a calamity. A calamity, according to Chinese proverb, is a time of great opportunity. Consuming nations are showing signs of reexamining their energy policy, as it is the time to give the pros and cons on the different substitute sources of energy (which is the subject of Chapter 3).

CHANGE IN THE MARKET STRUCTURE

"For . . . the major oil companies the good old days were those ten or twelve years which immediately followed the end of the Second World War."[48] Oil prices from World War II to the Suez War in 1956 were determined largely by the majors which could not only post a price but sell at that price. At that time nearly all OPEC oil was produced by the majors which maintained a monopolistic control over the world's oil prices. This was done through their ownership of affiliated refining and marketing companies in oil-importing countries.

Following the Suez War the last venture of two of the main colonial powers, Britain and France, in the Middle East was doomed. This also signaled the decline of other Western oil ventures. A few independents were now hunting for oil concessions in the Middle East. Although they were unable to obtain cheap concession rights, as the majors had obtained in the good old days, new patterns of financial arrangements were emerging, "for instance, the 75-25 principle whereby the foreign element had to put up the risk capital in the exploration stage and on discovery had to give 50 percent"[49] of the oil discovered to the indigenous partners. In the words of Adelman, a leading petroleum expert:

> The United States government, as trustee for the customers—and home government to most of the companies—compelled them to do what would not otherwise have happened, at least not so much so soon. After 1950, when the pressure disappeared, we see an actual reversal, with prices rising by perhaps 20 percent through 1956.[50]

The seven majors, or as their adversaries call them, "the seven sisters," were supported by the U.S. State Department in their oil policies.[51] The 20-30 "newcomers" operated, more or less, independently. While these independents found it difficult to market their products in the world as a result of the competition from the majors, they were successful in forging a breakthrough in the hitherto all-majors' preserve.

With the development of these independents or minors in international oil deals, national oil companies were also emerging in the producing countries. Usually, national oil companies try to export their oil in areas such as Eastern Europe where they are not competing with the majors. Their limited operations in the world market cannot yet be considered as a successful competition with the majors or minors in the world petroleum market. Yet these national companies emerged as "the children, so to speak, of public opinion and are viewed as symbols of the national aspiration to gain a place in the sun on the international scene."[52]

In the 1950s and also in the 1960s, the supply of oil exceeded demand. "The oil companies maintained below maximum-capacity levels on producing wells, held some wells in reserve, and left many proven fields undeveloped."[53] A classic example is the Iranian oil boycott initiated by the international oil companies. As soon as the boycott became effective the oil companies compensated the flow of oil to the world market by increasing production elsewhere. These were the days of the buyers' market. As the buyers' market was under the firm control of the multinational oil companies, the Arabs proved incapable of taking any firm measures against Western interests in their countries. The buyers' or consumers' market was undergoing radical changes with the increase in the growth of consumption of energy in the world. The market condition changed over the years. In 1925 oil accounted for 13.2 percent of total world primary energy consumption. This percentage grew to 29.7 in 1950, 37.2 in 1960, and 46.1 in 1970.[54] At the same time the contribution of solid fuels to energy was declining from 82.9 percent in 1925 to 59.3 percent in 1950, 46.5 percent in 1960, and 32.7 percent in 1970.[55] By 1973 oil increased its share by more than two-and-a-half times. By the same year world oil production had increased by more than four times.[56] To clarify the market picture, it would be of particular relevance to cite here the geographical distribution of oil production between the Western and Eastern Hemispheres. In 1950 the Western Hemisphere produced 71.9 percent of the total world oil production and in 1973 accounted for only 31.1 percent. While production was steadily declining in the West, it was increasing in the East. In 1950 the Eastern Hemisphere produced 28.1 percent of the world's total oil production, which increased to 68.9 percent in 1973.[57]

Although no trend can be discerned from the production figure alone, it is interesting to note that while oil production in the West was decreasing the demand for it was constantly increasing. The major industrial countries account for more than 75 percent of the world energy consumption. "Oil is the source of almost half of the energy in the United States and more than half in Western Europe and Japan . . ."[58] The increased demand for oil in the industrialized nations was being met by the supply of oil from the East. Following the Arab use of petrodiplomacy, the demand-supply ration was curtailed and an oil famine situation was created in the world market. Almost overnight the traditional buyers' market completely changed to a sellers' market. Germane to this context was the fact that the multinational oil companies, fearing loss at their business, represented the oil producers' interest and sided with them. Probably this was the sensible thing for the econopolitical multinational companies to do as they lacked support in their overseas operations from their own home governments which they used to receive in the past.

The price of oil, of course, does not so much depend on the normal working of the economic law of supply and demand. Over the decades oil has become more of an econostrategic commodity than merely a commercial product. Oil trade is more politically oriented than any other business enterprise in the world; there is also the fact that the oil resources are located mostly in preindustrial societies which, naturally, take a sense of psychic and political satisfaction in seeing the industrial giants throb. Added to these is the pattern of oil production and consumption. In many cases oil production is almost entirely controlled by the host governments. The multinational oil companies had little or no financial incentive to diversify sources of the oil regions; rather they concentrated on cheap sources of oil in a few areas of the world. Gradually, a near monopoly production of oil was developed in the Mideast, neglecting other areas of the world.

Because oil was one of the cheapest and a most convenient source of energy, the industrialized world changed its lifestyle from a coal-based source of energy to a depletable and highly politicized liquid. "At a time when oil was becoming more important to the countries that consume it,"[59] the producers were "seeking to industrialize their nations and build diversified economic infrastructures with oil revenues . . ."[60] For example, Saudi Arabia with the largest deposit of oil, which is also the principal commodity of the desert kingdom, was "tying oil supplies to joint-venture projects on their soil in an attempt to attract the kinds of energy-intensive industries that make use of their only resource."[61]

The immediate effects of the change in the market structure are that the price of oil is much higher now than it would have been in the normal market condition which prevailed in the world of oil before the Arab embargo of 1973. There is little basis upon which to predict any fixed oil price with a certain degree of accuracy. As discussed earlier, there is no possibility of depoliticizing the oil trade. All that can be said is that the market condition will remain unstabilized and the temptation for OPEC to increase the price either by a production cut or by bargaining will be great. In the past when "Iran and Saudi Arabia were submissive clients of the U.S.,"[62] and the Western powers were able to balance the energy equation by the sanction of force, the buyers enjoyed the use of cheap oil. "The energy crisis will stay with us after the political settlement of the Arab-Israeli dispute which has been the immediate cause of the present situation."[63]

The encouraging sign is that the bargaining leverage between oil exporters and importers is gradually shifting toward the consumers as they begin slowly to satisfy the demand for oil internally through substitution and other measures rather than by large imports. Since this is particularly true of the United States, a shift away from the sellers' to the buyers' market, with consumers leading the rescue operation, is conceivable in the not-too-distant future.

The unfortunate fact is that the vertiginous increase of the price of oil is no more a unique phenomenon. The price of fertilizer has risen almost as rapidly as that of crude oil. The price of foodgrain has tripled. "Some of the countries which have the least resilience in meeting the additional cost of oil imports are those which depend heavily on fertilizer imports"[64] and purchase of food abroad. If the prices of the three Fs (fuel, fertilizer, food) and other essentials continue to increase and nations continue to protect their individual or group interests at the expense of fellow nations, the economic consequences for the world may become severe.

NOTES

1. Christopher Tugendhat, Oil: The Biggest Business (New York: G. P. Putnam's Sons, 1968), p. 71.
2. Harvey O'Connor, World Crisis in Oil (New York: Monthly Review Press, 1962), p. 66.
3. Tugendhat, op. cit., p. 80.
4. Ibid., pp. 77-78.
5. Ibid., p. 93.
6. Edith T. Penrose, The Large International Firm in Developing Countries: The International Petroleum Industry (London: George Allen and Unwin, 1968), p. 62.
7. Paul H. Frankel, Essentials of Petroleum (London: Chapman and Hall, 1940), p. 91.
8. Ashraf Lufti, OPEC OIL, Middle East Oil Monographs, no. 6 (Beirut: Middle East Research and Publishing Center, 1968), p. 37.
9. Tugendhat, op. cit., p. 156.
10. Ibid., pp. 166-68.
11. Ibid., p. 165.
12. Helmut J. Frank, Crude Oil Price in The Middle East: A Study in Oligopolistic Behavior (New York: Frederick A. Praeger, 1967), p. 10.
13. J. E. Hartshorn, Politics of World Oil Economics (New York: Frederick A. Praeger, 1967), p. 343.
14. Joseph A. Yager and Eleanor B. Steinberg, Energy and U.S. Foreign Policy (Cambridge, Mass.: Ballinger Publishing Co., 1974), p. 14.
15. Ibid.
16. Roger LeRoy Miller, The Economics of Energy: What Went Wrong and How We Can Fix It (New York: William Morrow and Co., 1974), pp. 28-29.
17. Jahangir Amuzegar, "The Oil Story: Facts, Fiction and Fair Play," Foreign Affairs, July 1973, pp. 682-83. It may be noted that within three months of the publication of this article on oil facts, fiction, and fair play, the article became fiction as OPEC unilaterally violated the most important clause of the Tehran Agreement on collective bargaining.
18. Maureen S. Crandall, "Oil In The Middle East and North Africa," in Edward W. Erickson and Leonard Waverman, eds., The Energy Question: An International Failure of Policy, vol. 1 (Toronto: University of Toronto Press, 1974), p. 65.
19. Miller, op. cit., p. 29.

20. Taki Rifai, The Pricing of Crude Oil: Economic and Strategic Guidelines for an International Energy Policy (New York: Praeger, 1974), p. 263.

21. Petroleum Press Service, January 1973.

22. Neil H. Jacoby, Multinational Oil: A Study in Industrial Dynamics (New York: Macmillan Co., 1974), p. 259.

23. Ibid., p. 260.

24. Ibid.

25. Insight Team of the Sunday Times, Insight on the Middle East War (London: Andre Deutsh, 1974), pp. 178-78.

26. New York Times, October 18, 1973.

27. Insight Team of the Sunday Times, op. cit., pp. 178-79.

28. New York Times Magazine, March 24, 1974, p. 13.

29. Ibid.

30. Ralph H. Magnus, "Middle East Oil," Current History, February 1975, p. 50.

31. Raymond F. Mikesell et al., Foreign Investment in the Petroleum and Mineral Industries (Baltimore: Johns Hopkins University Press, 1971), p. 239.

32. Thomas R. Stauffer, "Oil Money and World Money: Conflict or Confluence?" Science 184, no. 4134 (April 19, 1974), p. 332.

33. Ibid.

34. Cited in Miller, op. cit., p. 30.

35. Washington Post, April 15, 1975.

36. See Edward J. Mitchell, ed., Dialogue on World Oil (Washington, D.C.: American Enterprise Institute for Public Policy Research, 1974), p. 46.

37. Amuzegar, op. cit., p. 689.

38. For an account of the shares of various national and international oil companies operating in Saudi Arabia, see Leonard Mosley, Power Play: Oil in the Middle East (Baltimore: Penguin Books, 1974), pp. 431-32.

39. Washington Post, February 5, 1974.

40. Washington Post, December 10, 1974.

41. Washington Post, December 14, 1974.

42. Ibid.

43. Elizabeth Monroe and Robert Mabro, Oil Producers and Consumers: Conflict or Cooperation (New York: American Universities Field Staff, Inc., 1974), p. 17.

44. Ibid., p. 22

45. Oil and Gas Journal, March 24, 1975.

46. Washington Star-News, December 13, 1974.

47. Cited in Monroe and Mabro, op. cit., p. 10.

48. Lufti, op. cit., p. 9.

49. Ibid., p. 10.

50. M. A. Adelman, The World Petroleum Market (Baltimore: Johns Hopkins University Press, 1972), p. 7.

51. Cited in Walter Laqueur, Confrontation: The Middle East and World Politics (New York: Bantam Books, 1974), p. 226.

52. Lufti, op. cit., p. 37.

53. Tareq Y. Ismael, The Middle East in World Politics: A Study in Contemporary International Relations (Syracuse: Syracuse University Press, 1974), p. 230.

54. Mahdi al-Bazzaz, "Middle East Oil Revenues: An Assessment of Their Size and Uses," Middle East Economic Digest, March 15, 1974.

55. Ibid.

56. Ibid.

57. Ibid.

58. S. David Freeman, Energy: The New Era (New York: Vintage Books, 1974), p. 5.

59. Ismael, op. cit., p. 233.

60. Ibid.

61. Ibid., p. 234.

62. Laqueur, op. cit., p. 229.

63. Middle East Economic Digest, December 7, 1973.

64. International Bank for Reconstruction and Development and International Development Association, Additional External Capital Requirements of Developing Countries (Washington, D. C., March 1974), p. 2.

CHAPTER

3

SIGNIFICANCE OF ALTERNATE SOURCES OF ENERGY

In 1973, an eventful year for the oil industry, the consuming countries experienced a tightening supply-demand relationship. The explanation lies in the fact that 75 percent of the world oil reserves and over 46 percent of the supplies are derived from the Middle East. The basic reason for the problem was that the demand for petroleum products in the United States increased steadily over the previous decade while domestic supply declined. The reason why the United States is less self-contained today than ever before lies, first, in the changed picture of international politics and, second, in the Arab boycott of oil shipments. In a divided world, when the Russians had practically no foothold in the Middle East, the United States has had leadership thrust upon it. The supply source of such strategic material as oil was considered guaranteed. Following the development of detente and a considerable Russian influence in the Middle East in 1973, the Arabs had added encouragement to impose an embargo on sales of oil to countries which followed a pro-Israeli policy. In political terms, the Arab oil embargo of 1973-74 made the United States and other industrialized nations of the non-Communist world conscious of the unfriendliness and unreliability of the Arab countries.

The oil-exporting countries have been worried for some time about what would happen when they ran out of their finite resources—oil. Some of the members of the Organization of Arab Petroleum Exporting Countries (OAPEC) have small populations and huge accumulated cash, more than they can spend. Consequently, they are able to hold back production at will and in a crisis situation aggravate it further by embargoing supplies to the leading consumers for political reasons. Once they have used the oil weapon successfully they would have little reason not to use it again. Speaking to the American Society of Newspaper Editors in Washington on April 18, 1975, Saudi Oil Minister Yamani left open the possibility that there could be a new oil embargo if peace is not achieved in the Middle East.[1] The Arab-Israeli conflict is a political problem and bears little or no relevance to the energy supply-demand situation in the world. But the Arabs seem to be determined to wield the oil weapon again to accomplish their political objective. There are many temporary remedies to the energy crisis which will be discussed later, but the ultimate solution of it lies in the development of alternative sources of oil.

The United States is both the largest producer of oil and its largest importer. The U.S. dollar is drained abroad by the purchase of additional oil at higher prices. The American oil import bill skyrocketed from $8.3 billion in 1973 to $25 billion in 1974.[2] Inflation in the United States reached about 12 percent in 1974, while inflation in most of the rest of the free world about doubled that rate. Like its dollar, hardship in the United States is automatically exported, and the world then suffers more than does the United States. The Arab oil cutoff in 1967 did not succeed because the United States was able to resist the pressure and could help its European allies by means of the oil supply then available in the United States. The same action could not be taken in the winter of 1973-74, inasmuch as the United States itself was then dependent on Arab source of oil supply.

In an international financial sense, U.S. money is not a national currency, but it is the international currency. Transfer of too much American money to a group of oil producers is not only a danger to the health of the U.S. economy, but it is proving to be more dangerous to the world financial adjustment.

According to a World Bank estimate, if oil importers maintain their present purchase rates, OPEC nations will amass a currency reserve of some $653 billion by 1980, which is three times the total present currency reserve of all the nations in the world.[3] At present a greater amount of the petrodollar is being recirculated in the oil-importing nations through bank deposits and purchase of weapons and goods. The possibility remains that if there is another Arab-Israeli war Arab nations, which will by then have the lion's share of the currency reserve, might suddenly withdraw their oil money. The inherent danger in this action would be near strangulation of the Western economy.

Not only the United States but also other industrialized non-Communist nations and the Fourth World countries found in the Arab oil embargo an increasing danger to their economic growth and standard of living. For reasons of national security the United States is particularly alarmed by its growing dependence on imported oil which may again be subject to manipulation of supplies for political purposes to exploit American vulnerability. The Arab oil embargo has provided the United States with the specter it needs to muster public support for its Project Independence. A consensus is growing that the United States should launch a unified, vigorous research and development effort into alternative energy sources. The United States has technology and resources in a manner not possessed by any other nation. Given the time and money there is little doubt that the United States will develop alternative energy sources. Needless to say, U.S. energy self-sufficiency is important not only to prevent oil embargo against it and its allies, but to save the oil-short and oilless nations of the Fourth World from economic collapse as well. Here lies the significance of leadership for the development of alternative energy resources.

The postembargo period has set oil-importing nations on a course which demands self-sufficiency in energy by means of substitution. In fact, it may have been a blessing in disguise for the world that following the oil embargo the non-Arab nations took the lead in raising the price of oil to a level high enough to make it competitive with most alternative inexhaustible sources of energy. A cut in oil prices would probably reverse the situation by making alternative sources of energy less competitive with oil. But, judging from the sufferings that the industrial world has had in the postembargo years, a complete reversal of the process is not expected. However, if the oil-exporting nations, after accumulating huge sums of money, decide on their own to reduce drastically the price of oil, making substitution less attractive, the industrial world could very well fall again into the trap of the Third World. Should this happen, the consequences would be devastating for the whole world when oil resources are depleted. An energy doomsday prophecy could become a reality. Ironically, to prevent this energy doomsday it is the oil-exporting exporting nations which could do well not to reduce the price of oil very much and to invest their petrodollars in the development of alternative sources of energy in the West.

Of all the industries in the world, that of oil leads in volume and monetary value. It is the one industry that concerns every country in the world and affects the lifestyles of almost all living beings in varying degrees. Oil has revolutionized civilization and continues to energize it. A day rarely passes anymore that worldwide media do not headline and emphasize the oil situation. Why? There are many reasons: The bulk of the oil trade moves internationally; huge profits are made by the exporting countries and oil companies; the oil industry is vital to producing and consuming nations alike; oil import affects the balance of payments of both the developed and the developing nations; oil is a vital necessity; and the lack of oil could paralyze our modern civilization. Inevitably, because of these and other attributes, oil is enmeshed in international political and economic pressures among various contenders—companies, governments, and regional and universal international organizations. Thus, oil is presently a commodity like no other and is supremely an econopolitical industry.

The contrasting and, in some ways, conflicting interests of the producing and consuming nations and the international oil companies have gained the attention of the governments and peoples of the world. Because of the actual embargo of oil an international public opinion has been created favoring displacement of oil by secure energy sources within national boundaries. Added to this is the higher prices of oil which are forcing the energy consuming world, particularly the United

States, to search for new sources of energy. The events that followed the 1973 oil embargo have substantially modified previous trends in energy consumption. The new trend is to save energy through conservation, efficient use of oil, and substitution, which is becoming the norm now.

With this background on the significance of alternative energy, an attempt will now be made to evaluate the economic and political factors that will play a part in determining the present and future sources of substitute energy in the world. Technological developments in energy production affect the energy consumption, mainly by reducing costs in the production and transportation of primary energy sources and in their transformation into secondary energy. This discussion will cover both aspects of energy. The ways of the energy development are complex: Cause-effect relationship often cannot be determined unambiguously. The voice and rhetoric of the political leadership is but one, and sometimes not even a single one, in a chorus. However, as it is, energy can be obtained from different alternative sources: It can be derived from coal, petroleum, gas, oil shale, tar sand, hydropower, nuclear energy, uranium, geothermal and solar energy, tidal power, and wind power.

Under normal conditions, one would expect that, as oil and gas prices increased in proportion to their scarcity, the more abundant and less developed fuels would appear in the marketplace as substitutes. But the approach to date to achieve a commercial status for those processes that involve the derivation of liquid fuels from unconventional sources in place of crude oil has met with slight success, due to a number of logistical, technical, and economic reasons.

For many years, particularly since the Suez crisis of 1956, the governments of countries in which oil imports are of major importance have been less conscious of the problem posed by the vulnerability of their sources of supply. The United States, which fulfilled its contingent liability as a supplier of oil to Western Europe during the 1967 Arab-Israeli war, found itself dependent on imports from the Middle East during the 1973-74 oil embargo. Project Independence provides a framework for a massive effort by the U. S. government for developing a national energy policy to free the American consumers from heavy dependence on foreign sources of oil supply. This is an American program to eliminate imports of oil by developing substitute and through discovery and recovery of oil in the United States. Project interdependence, on the other hand, implies energy sharing, stockpiling, conservation, international financial cooperation, research and development through collective effort by the members of the International Energy Agency (IEA) which was founded in 1974 at the initiative of the United States. The IEA membership includes almost all Western industrialized nations and Japan. A plan for energy independence, on Project Interdependence based in the first instance on American national interest, could be an essential ingredient in a new approach to the energy situation by the whole world. Project Independence "could make a vast difference to international relations in the next 10 to 20 years, and serve as a step toward a more rational world use of energy for the benefits of man."[4]

Project Independence is a novel experiment on the part of the U.S. government in trying to meet some extremely difficult goals in a relatively short time. The objective is not to have zero imports by 1980. It is to remove the dangerous degree of dependency on imported oil. Traditionally, the energy industries have been the preserve of private enterprise in the United States. The consuming sectors, consisting of private citizens, commerce, and industry, involve the public at large. Ultimately it is their actions which will determine to what extent growth in energy demand can be restrained.

Sources of energy are judged to be potentially significant where the available quantities are large and where technological and economic considerations show that the costs are competitive. Enough work has been done to permit a reasonable assessment of the major sources of alternative energy.

TIMETABLE FOR DEVELOPMENT

The central thesis of this study is that, in the long range, oil cannot be used as a weapon of diplomacy, since it is likely to be replaced by effective substitute energy. A companion thesis is that, in the short range, the Arab use of oil as a weapon of diplomacy has a limited scope in the settlement of the Arab-Israeli conflict. Putting these theses into perspective calls for an evaluation of the three time frames seen for developing energy independence: 1975-85, 1985-95, and 1995-2005. Evaluating the strengths and weaknesses of the time frames requires the review of a few key features of the short-term measures directed toward energy self-sufficiency in the United States, inasmuch as any potential Arab oil embargo will be directed primarily against the United States.

Many short-term measures seem to be viable alternatives for dangerous dependency on Arab oil. However, only the most feasible alternatives will be reviewed here. The present oil crisis is likely to shake the world for about five years. By 1978, oil will be flowing in from the Alaska pipeline at the rate of 600,000 barrels per day, but even the flow of 2 million barrels a day by 1981 will not free the United States from the need for foreign supplies. At about the same time, Britain and Norway each expects to get 2 million barrels a day from the North Sea. The United States could expand its domestic supply of oil by increasing the capacity of the Alaska pipeline and by opening the navy's petroleum reserve in California and Alaska. Alaska's north slope, known as Naval Petroleum Reserve No. 4, is believed to contain as many as 33 billion barrels.

The United States could encourage more offshore drilling, and help enhance the exploration, production, and utilization of oil resources in other parts of the world. A most encouraging sign is that there is a scramble in the search for oil worldwide. The prospect for discovering

MAP 3

The Worldwide Search for Oil

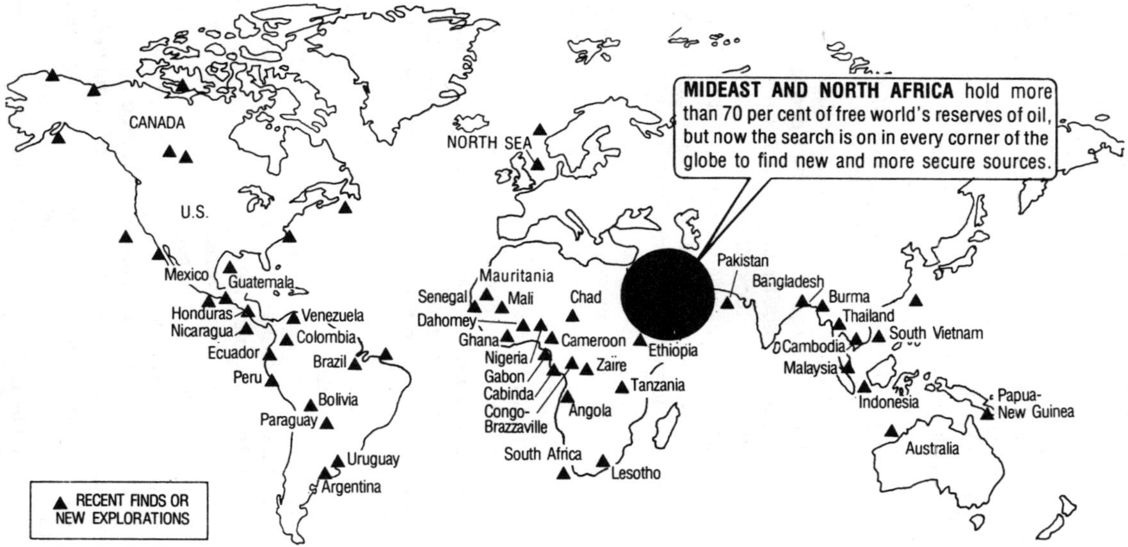

Source: U.S. News & World Report, December 30, 1974.

oil in the fuel-deficit areas of the world may alter the global energy picture as it is known today.

An interim measure of stockpiling and conserving energy is essential during the 1970s. Simultaneously, it is necessary to maintain diversified sources of foreign supply of oil to the United States. For a specifically limited time, the lowering of pollution standards on the burning of coal and oil, and the elimination of pollution devices on automobiles may be necessary to bolster energy independence programs in the United States. While no single dramatic breakthrough is expected, ways to cut waste in energy consumption, improvement in gasoline mileage, introduction of an ultrasonic reactor that mixes oil and water in the auto engine burner thereby reducing consumption of oil, and the innovation of nuclear-powered transportation either through electrical transportation or through hydrogen fuel—a combination of all of these measures—could reduce gasoline consumption in the United States.

Energy experts at the Massachusetts Institute of Technology (MIT) in 1974 recommended that the United States build up a stockpile of crude oil to offset a possible oil embargo. Their study, published in book form, said that a combination of stockpiling and other measures would be the best way to reduce embargo risks.[5] The need for stockpiling is urgent for the United States, since it alone among the major industrial consumers has no provision for stockpiling crude oil beyond normal inventories.[6] The MIT group believes that if the United States required oil companies to maintain a stockpile of 2 million barrels a day the cost to consumers would rise no more than 25 cents per barrel or about two-thirds of a cent per gallon.[7]

Stockpiling of oil has been successfully adopted by the European countries, and there is no reason to believe that it cannot be done effectively in the United States. The size of the stockpile depends on the magnitude of an expected oil embargo, but reserve for one year's consumption of foreign oil may be enough to cushion the shock of future emergency. A reserve of 1 billion barrels would permit America to draw 3 million barrels a day for one year in case of another embargo. The oil stockpile, however, has some pitfalls, unless accomplished through conservation. If attempted by increasing imports, the price of oil could go up even further in market competition, or the producers might be tempted to raise prices again, either by cutting production while demand is up or by simply increasing the price of their monopoly export. On the other hand, maintaining the present level of imports and conserving 3 million barrels a day through such measures as were practiced in the winter of 1974 may not have any adverse effect on price and the American economy, but those were emergency measures taken as a result of the serious oil shortage during the embargo. More than likely it will be quite difficult during more normal times to convince Americans that their century-long wasteful habits of energy consumption must be changed.

Many studies have been published showing that the United States can cut back its oil consumption by 2.5 million barrels a day—more than 15 percent—without much hardship.[8] A most noteworthy study on energy conservation is the Ford Foundation's $4 million inquiry into the American energy crisis. The authors—a team of economists, lawyers, and scientists, among others—maintain that the United States can balance its energy budget and avoid reliance on insecure oil sources abroad by slowing its growth rate in energy consumption.[9] The authors believe that American energy growth can be trimmed to about 2 percent a year or, in time, even to zero without adversely affecting the economy.[10]

It was through conservation measures that oil consumption fell drastically in the first half of 1974 in the European Common Market countries. Oil use fell by 15.5 percent in West Germany, 9.9 percent in Britain, and 6.1 percent in France, compared with the first six months of 1973. Consumption dropped even more in some of the smaller countries—22.8 percent in the Netherlands, 19.1 percent in Belgium, 15.7 percent in Denmark, and 16 percent in Luxembourg.[11]

While stockpiling and conservation programs should go hand in hand, diversification measures of foreign sources of oil supply to the United States must be followed. During the first eight months of 1974, Canada and Nigeria combined exported to the United States more than twice as much crude oil as all of the Arab countries put together. The largest Arab oil exporter to the United States was Saudi Arabia, which occupied fifth place in the total picture, as Table 7 shows.

TABLE 7

Oil Exports to the United States, January-August 1974

Exporting Country	Thousands of Barrels	Percent of U.S. Imports
Canada	896.4	24.7
Nigeria	633.8	17.5
Iran	561.9	15.5
Venezuela	405.1	11.2
Saudi Arabia	301.0	8.3
Indonesia	283.7	7.8
Algeria	195.7	5.4
Ecuador	78.7	2.2
Trinidad	78.6	2.2
United Arab Emirates	62.6	1.7
Angola	47.3	1.3
Others	80.1	2.2
Total	3,624.9	100.0

Source: Federal Energy Administration figures in the *Washington Post*, October 24, 1974.

Production of domestic oil within the continental 48 states has been declining since 1970, and the flow of oil within the United States, except for the state of Alaska and possibly Hawaii, may fall even further. It may be that the only way to reverse the trend is to step up offshore explorations in the mid-Atlantic, the Gulf of Alaska, Cook Inlet in Alaska, and the Santa Barbara Channel off the coast of southern California. These and many other proposals are parts of America's Project Independence programs to achieve self-sufficiency in energy. The attempt here to describe the three decade-long time frames for development of "nucosolar" (nuclear, coal, and solar) energy is, of necessity, mutually inclusive with all the conventional sources of energy along with stockpiling and conservation programs to tackle any crisis situation in the event of another embargo during the 1970s.

The timetable given for the expanded energy availability varies from person to person and study to study. The consensus is that there are three elements involved in working out a time frame for probable new future energy development. The elements are political, economic, and technological. Disagreement is narrowing and there is emerging a convergence of opinion that what is feasible technically should receive political and economic support of the leadership. President Ford's energy message in his State of The Union message in January of 1975 and the Democratic party's lead in announcing its energy programs suggested such a convergence of opinion. The seriousness of these messages implied that the era of cheap energy is gone, possibly forever. The price of oil may come down a little due to market pressure or the possible breakup of OPEC, but never again would it go as low as it was in 1973. In a sense the higher oil price is essential in order to develop alternative renewable sources of energy, otherwise the world would depend on the same cheap, but nonrenewable, source of oil,

President Ford's first budget proposal announced in February 1975 gives due importance by making adequate provisions for increased expenditures for development of alternative sources of energy. For development of nuclear energy research, budget request is for $1.634 billion, an increase of $313 million. The allocation for further research and development of thermonuclear fusion rises to $120 million from $85 million. The fund for fossil fuel research is over $340 million, an increase of $130 million. The most dramatic increases in the budget proposal are for development of solar and geothermal energies. The allocations of fund for these newer sources of energy are $91 million, a rise of 727 percent.[12]

Enough study has been done to permit a reasonable assessment of the energy sources that today appear novel. The first in the series to be considered is the production of electricity from nuclear fuels. In any discussion of nuclear fuel the public mind reacts as though to the danger of the atomic bomb. As many as 60 independent scientists have concluded in a study done for the reorganized U.S. Atomic Energy Commission (AEC) that worries about nuclear danger have been vastly exaggerated.[13] James T. Ramey, a former commissioner of AEC, maintains that today's reactors feature enough built-in safeguards to prevent accidents.[14] Nuclear power is two-faced; it is clear and does not pollute the environment as fossil fuel does, but its radioactive waste is dangerous for many lifetimes. The social cost of nuclear power is being measured; the U.S. federal government has invested over $1 billion in an attempt to measure environmental and societal costs associated with nuclear power.[15] As a result, a vast amount of literature has appeared providing material for public debate on nuclear hazards. The debate is unbalanced in the absence of any parallel assessment of the danger of burning fossil fuels. The issue can hardly fail to be resolved as more studies on the danger of burning fossil fuels appear. On balance, David J. Rose, a nuclear scientist, concludes:

> The uranium and thorium resources, the technology and the societal impacts all seem to presage an even sharper increase in nuclear power for electric generation than had hitherto been predicted.[16]

The AEC predicts that 132,000 megawatts of nuclear power will be available in the United States by 1980. The Federal Power Commission (FPC) predicts that only 122,000 megawatts will be on line by that time. Taking the conservative estimate of FPC, it is reasonable to say that one-third of America's electricity will be generated from nuclear fission in 1980.[17] By 1985 the electrical power generation from nuclear energy will expand to 42 percent.[18] As electricity consumption grows, the contribution of nuclear power will undoubtedly grow as well.

Ramey, giving an historical review of the U.S. nuclear power program and discussing its present contributions and future prospects,[19] said that the 38 thermal (light-water) reactors now under operation in the United States, the large number of them under construction and planned, and the breeder and fusion reactor concept being satisfactorily developed give indications that nuclear energy will be a safe, reliable, and economic source for generating electricity.

Ramey predicts:

> In the years ahead, we will see great technological advance throughout the world and, hopefully, corresponding social and economic progress. The atom—by providing a virtually inexhaustible resource for low cost energy and by serving mankind in countless other ways—can help us achieve this promise for the future.[20]

Short of a major accident in the nuclear power plant that could force political reconsideration, nuclear power will continue to grow rapidly in the United States as well as in Europe. The forecast of uranium shortage can impel a switch to breeder reactors, when the production in

MAP 4

The Nuclear Power Grid
(States sized according to their populations)

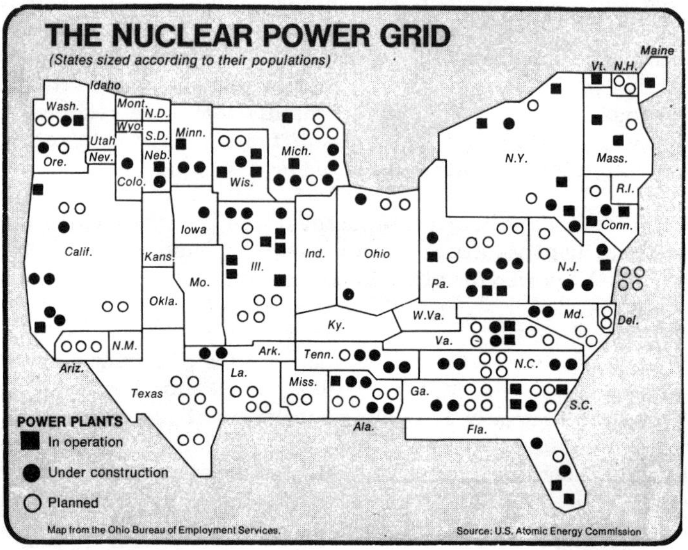

Source: Cited in Newsweek, December 10, 1973.

thermal reactors has peaked after 1985. The large-scale use of nuclear power in the 1980s will release much needed oil and natural gas for other pressing uses, particularly for automobiles.

An energy-autonomous United States can retain its own capacity to counter oil embargo by helping the Western European nations against their vulnerability. A European partial energy independence program can only be based on nuclear development, since the continent lacks oil resources. The American action in reorganizing the AEC under two new agencies—the Energy Research and Development Administration (ERDA) and the Nuclear Regulatory Commission (NRC)—is in the right direction, the direction to find better ways to develop nuclear power and establish public confidence in its safety. It is in this field that nations on both sides of the Atlantic may cooperate to overcome the challenge and the hitherto insurmountable technical problems in the nuclear arena. West

European nations, including "pro-Arab" France, are drawing closer to the United States on energy matters. Japan, although a developed nation, is physically located in the less-developed region of the world. Already Japan has drawn closer to the United States and the oilless poorer nations in Asia, Africa, and Latin America are likely to seek cooperation with the United States to draw the benefits of technical development in peaceful uses of nuclear power. Egypt is an oil-short country and it is coming closer to the United States and France by getting nuclear reactors from both countries.

In the fields of energy and related matters American records of innovation are unsurpassed. Oil was first drilled and exploited on a large scale in the United States. The large-scale use of coal and steam to operate factories and trains was also an American first. Electricity was invented in the United States. Mass production of automobiles is also an American contribution to modern man's independent mobility. The atomic bomb, space exploration, and the landing on the moon are but a few examples of American scientific feats, leading the world.

Many American scientists believe that the fuel of the future is hydrogen. An innovation in this field is the use of hydrogen to fuel the internal-combustion engine. American astronauts have used hydrogen power cells for electrical power in their spacecraft. Now, scientists at the Jet Propulsion Laboratory in Pasadena, California, have successfully demonstrated an automobile engine that runs on a mixture of hydrogen and gasoline. Its operation is clean, meeting almost all of the federal emission standards set for 1977-model cars.[21]

Another group of American scientists has developed an engine that converts all its gasoline into hydrogen. Pilot studies are also being conducted to provide pure hydrogen for propulsion.[22] These innovations, along with the attempt to use hydrogen for heating and cooling homes, may not only bail America out of the energy shortage, but save the world from future economic peril. Hydrogen-powered automobiles could be mass-produced in ten years or less.

There are a variety of alternative programs envisioned by energy experts in the United States to reduce or eliminate dependence on external sources of expensive oil. They even include garbage-power-conversion of wastes into oil—and all the other competitive sources of energy referred to earlier in this study. Our second selection for possible development within the timetable set by us include synthetic crude oil (syncrude) and synthetic gas (syngas). This selection, like the first one, nuclear energy, is not made arbitrarily but is based on technical feasibility and resource availability with financial consequences that the American economy can bear.

The United States has 3.2 trillion tons of coal deposits; therefore, with the proven oil deposit running out fast, coal seems to hold the promise for a fuel-based transportation system. Syncrude, syngas, and methyl alcohol (methanol) can be produced from coal. The advantage of coal gasification and liquefaction has been described by Wilson Clark in the following work:

> Besides the convenience offered by gasification and liquefaction in terms of not having to convert or replace existing fuel-burning devices to utilize them, these two related technologies also offer another attractive advantage over coal in its natural state; and that is that the pollutants—the sulfur, the nitrogen oxides, the ash, etc.—are removed in the gasification process, leaving a clean fuel. It is both cheaper and simpler to remove pollutants at this stage than to try to remove them from the smokestack after coal is burned as a fuel.[23]

Although there is great uncertainty in predicting the costs of the new technology, the MIT Energy Laboratory Policy Study Group notes that for all processes involved in this new technology the capital costs are very similar to that of the old technology.[24] The National Petroleum Council (NPC) has projected that by 1985 syngas will supply 2.5 trillion cubic feet of gas per year "at the maximum rate physically possible without any restrictions due to environmental problems, economics, etc."[25]

The NPC predicts that the demand for gas in 1985 will be 41 trillion cubic feet; the supply of domestic gas will fall short by 10 trillion cubic feet.[26] To fill this gap with syngas would require the construction of a minimum of 120 coal gasification plants before 1985 at a capital cost of $24 billion.

Research and development for syncrude is slower than for syngas. By 1985, the United States could have 10 syncrude plants, each with a 100,000-barrel-per-day capacity.[27] The oil shortage in the United States has sparked new interest in coal liquefaction technology, since coal is the largest energy resource of the United States. By producing syncrude from coal the Americans could maintain a 150-year supply.[28] This is in addition to coal's other uses. Thereafter, if oil will have any use it may be produced from the shale deposits in the United States and tar sands in Canada. But oil may become worthless long before the United States runs out of its coal deposits. Taking into consideration the rise in the cost of imported oil, syncrude is economically competitive now.

The editor of *Science* magazine, in an editorial, urges increased federal funding for coal liquefaction:

> In comparison with the billions we spend on oil imports . . . the millions the government is devoting to liquefaction of coal can best be described as a phony commitment—a cosmetic effort whose purpose is to give the appearance, but not the reality, of action. A goal worthy of the world's leader in technology would be to construct in 2 years several plants, each costing about $1 billion and each capable of supplying one percent of liquid hydrocarbons we consume.[29]

According to Yale University economics Professor William D. Nordhaus, the cost of continued reliance on foreign oil during the two decades from 1975 to 1995 would come to $663 billion, while the total for energy independence would amount to $985 billion, a difference of $16 billion per year.[30] This is not an impossible task for the United States. If further exploration and development of natural gas and oil from the outer continental shelf does not cut oil imports to zero by 1985, which is unlikely, the United States may not have any other alternative but to invest the $16 billion additional amount of money every year to achieve overall energy autonomy.

The foundation for development of syngas and syncrude will be firmly laid by 1985. Although no study or data are available now it may be safely assumed that during the decade 1985-95, syngas and syncrude will replace natural gas and oil. Coal was king and is an old reliable abundant source of energy for the United States. Syngas was once widely used in the United States.[31] It was about 1955 when the utilities in the United States shifted from syngas to natural gas. The process can be reversed without hardship on the economy or on the consumers. Similarly, the new technology of developing syncrude can be largely functionalized without major hardship within a decade.

The energy problems facing the world in general and the United States in particular do not require exploitation of every energy source as there are so many of them amenable to technical solutions. Our efforts have been to concentrate on the most feasible ones—technologically, financially, politically, and environmentally. There are several forms of energy that are eternal in nature. These are thermonuclear fusion, and solar and geothermal energies (tidal and wind powers are not considered here as they are dependent on solar power). Among them we have already covered fusion power, syngas, and syncrude. The probability of exploiting solar energy during the decade 1995-2005 will be our main concern now to the exclusion of geothermal energy, as the development of two eternal sources of energy may not be required. Another reason for our exclusion of geothermal energy is that simultaneous development of two parallel eternal sources of energy is not a financially sound proposition. Finally, we have excluded geothermal energy considering the fact that "Theoretically, as the earth's heat is converted into electrical power, the total heat in the earth would drop."[32] Thus in the strictest sense the only inexhaustible primary energy is radiated from the sun.

Many prestigious scientists and engineers have begun working on methcds for converting solar energy into heat, electricity, and chemical fuels:

> Within 5 years, many of these scientists believe solar-powered systems for heating and cooling homes could be commercially available at prices competitive with gas or oil furnaces and electric air conditioners. More significant, but

> farther in the future, is using heat from the sun to generate electricity; experimental solar-thermal generators have been constructed in several countries, and several groups in the United States are designing systems that take advantage of improved materials and manufacturing techniques. Eventually the direct conversion of solar radiation to electricity by means of photovoltaic cells or its bioconversion to wood, methane, or or other fuels on a large scale may become economically feasible.[33]

Solar water heaters have been in commercial use in California and Florida since the 1930s.[34] They are also in use in many European countries, Japan, the Soviet Union, Australia, and Israel. "The heating and cooling of buildings by solar energy is an early practical and economic probability."[35] Their widespread use in homes is expected after 1985.[36] Solar radiation is so abundant that the energy arriving from the sun on 0.5 percent of the land area in the United States is more than enough of the total energy requirement of the United States projected for 2000.[37]

Because of the cost and difficulties of installing solar energy in existing homes and because of the slow rate of replacing housing, the technology of solar heating and cooling will take several decades before it could have any significant impact on energy use. There are various projections on this. The Ford Foundation's Energy Policy Project estimates that it would take a half century from 1975 for the impact of solar energy on the society to be felt, at which time new housing will replace existing homes.[38] Scientists at the American Association for the Advancement of Science feel that because of the growing shortage of fuels solar energy will be widely used within several decades.[39] As society begins to realize that cheap fossil fuels are not easily available, scientists have suggested that collection of solar energy on house rooftops may be explored. This process is already in use here and there in the United States. Its widespread use in existing homes is a real possibility as conventional energy sources are dwindled and as the political climate hitherto inimical to development of solar energy disappears. Interest in large-scale application of solar energy has been accelerated in the United States. Both former President Nixon's and President Ford's energy messages pointed to the direction of fast development of solar energy.

In prospect, we can refer here to a conclusion reached by a professor of engineering on the development of solar energy as follows:

> In favorable locations, the costs of solar heating and cooling equipment under development appear to be nearly competitive with fuels; hence, this application is expected to be widespread within a very few years. Electric power from solar energy is not now competitive with conventional supply and is, therefore, a longer term possibility.[40]

We have considered various shadings of expert opinions on solar energy. It seems that the development of solar electricity will take more time. It may not even be necessary to explore the feasibility of solar electricity if the nuclear breeder and fusion reactors are successfully developed. The development of solar heating and cooling is an early possibility. Their large-scale utilization from 1995 is a practical and realistic projection based on highly sophisticated studies and expert opinions. As the new century dawns, the saga of solar houses may add a new chapter in the lifestyle of twenty-first century man.

In retrospect, it seems that Saudi Arabia, today's largest exporter and reservist of oil, is the greatest beneficiary of the oil industry that the Americans pioneered. Saudi Arabia not only has the largest deposit of oil but the solar energy that radiates on its desert each year is plentiful and about equal to the world's entire proved reserves of coal, oil, and gas.[41] For its part the neocapitalist Saudi Arabia has shown some interests in investing its petrodollars in the West to develop energy supplies from the sun. There was a contention prior to Faisal's death that "Faisal does not want to bring down prices [of oil] now and throw away his bargaining power for a settlement with Israel."[42] The higher price of oil may be a blessing in disguise for the world, since solar energy is competitive now with fossil fuels. As the reserve of fossil fuels begins to deplete in the world, it is high time to develop alternate eternal sources of energy. Saudi investment in solar energy, particularly in solar cooling and electricity, will go a long way to benefit her when the oil resource of the kingdom is exhausted in about 2039, if the present rate of production continues. As Saudi Arabia is the greatest beneficiary of the development of the oil industry, so is it also potentially one of the greatest beneficiaries of commercialization of solar energy, since the desert kingdom has the largest concentration of solar radiation on its territory.

Summing up, we can say that oil will surely maintain its favored position as a source of primary energy during the initial years of the decade 1975-85 and will decline progressively thereafter. The crucial period is the first five years of the decade as the effect of the steady shift away from oil toward the use of nucosolar energy begins to be felt by 1980 and the years following.

The eternal power sources are technologically beyond reach now, fossil fuels of oil and gas are running short, and coal gasification is still not widespread. Looking at the technological and economic possibilities of all alternative sources of energy that can be greatly expanded in the future, we see that these are nuclear fission, gasification and liquefaction of coal to produce syngas and syncrude, and solar energy. We should not, however, think of a nucosolar energy future only. "Rather, we should expect to be using an energy mix, just as we do now, with each energy source supplying the requirements which it can satisfy in the most suitable way . . ."[43] But nucosolar energy is likely to play the most significant role in three overlapping decade-long time frames, beginning in 1975.

The most dramatic changes projected to take place during the first and second decades are in nuclear power development and the use of syngas and syncrude respectively. The next immediate probability is the large-scale use of solar energy. Technical feasibility for all of these alternatives has been established. Based on current oil prices the economic-competitive compatibility of nucosolar energy, except for sun electricity, is already well recognized. To maintain this economic competitiveness of nucosolar energy and other possible substitutions, such as oil shale, oil sands (tar sands), geothermal energy, etc., higher prices of oil are essential to motivate the fast development of alternative sources of energy through private investment in the Western world, particularly in the United States.

NOTES

1. Washington Post, April 19, 1975.
2. Melvin R. Laird, "Let's Meet The Energy Crunch Now," Reader's Digest, January 1975, p. 50.
3. Cited in Ibid.
4. Carroll L. Wilson, "A Plan for Energy Independence," Foreign Affairs 51, no. 4 (July 1973), p. 675.
5. MIT Energy Laboratory Policy Study Group, Energy Self-Sufficiency: An Economic Evaluation (Washington, D.C.: American Enterprise Institute for Public Policy Research, 1974).
6. Ibid., p. 69.
7. Ibid., p. 70.
8. See the report of the press conference by Senator Frank Church, chairman of the Senate Foreign Relations Subcommittee on Multinational Corporations, as published in the Washington Post, January 12, 1975.
9. Energy Policy Project of the Ford Foundation, A Time to Choose: America's Energy Future (Cambridge, Mass.: Ballinger Publishing Co., 1974), pp. 45-79.
10. Ibid., pp. 81-111.
11. Washington Star-News, December 18, 1974.
12. Budget of the United States Government, Fiscal Year 1976, Appendix (Washington, D.C.: U.S. Government Printing Office, 1975), pp. 755-762; Washington Post, February 4, 1975.
13. See Laird, loc. cit.
14. The World Almanac (New York: Newspaper Enterprise Association, Inc., 1975), p. 114.
15. David J. Rose, "Energy Policy in the U.S.," Scientific American 230, no. 1 (January 1974), p. 359.
16. Ibid.
17. "FPC: 30 Nuclear Plants Delayed," Info, no. 30 (March 1973), p. 4.

18. National Petroleum Council, *U.S. Energy Outlook: An Interim Report* (Washington, D.C.: National Petroleum Council, 1972), p. 20; *Newsweek*, December 10, 1973.

19. James T. Ramey, "The Promise of Nuclear Energy," *The Annals*, November 1973, pp. 11-23.

20. Ibid., p. 23.

21. *Newsweek*, November 12, 1973.

22. Ibid.

23. Wilson Clark, *Energy for Survival: The Alternative to Extinction* (Garden City, N.Y.: Anchor Books, 1974), p. 254.

24. MIT Energy Laboratory Policy Study Group, op. cit., pp. 52-53.

25. National Petroleum Council, op. cit., pp. 17, 44.

26. Ibid., p. 37.

27. Lawrence Rocks and Richard P. Runyon, *The Energy Crisis* (New York: Crown Publishers, 1972), p. 40.

28. Ibid., p. 39.

29. *Science*, June 15, 1973.

30. *The World Almanac*, op. cit., p. 115.

31. Allen L. Hammond et al., *Energy and The Future* (Washington, D.C.: American Association for the Advancement of Science, 1973), p. 12.

32. Rocks and Runyon, op. cit., p. 57.

33. Hammond et al., op. cit., p. 61.

34. Clark, op. cit., p. 370.

35. George O. G. Lof, "Solar Energy: An Infinite Source of Clean Energy," *The Annals*, November 1973, p. 56.

36. Energy Policy Project of the Ford Foundation, op. cit., p. 51.

37. Hammond et al., loc. cit.

38. Energy Policy Project of the Ford Foundation, loc. cit.

39. Hammond, et al., op. cit., pp. 62-63.

40. Lof, op. cit., p. 52.

41. Energy Policy Project of the Ford Foundation, op. cit., p. 313.

42. *Time*, January 6, 1975.

43. Martin Wolf, "Solar Energy Utilization by Physical Methods," *Science* 184, no. 4134 (April 19, 1974), p. 386.

CHAPTER

4

THE ARAB-ISRAELI CONFLICT AND SAUDI POLICY

Saudi Arabia has been supporting the Palestinian Arab claims at almost every opportunity. King Ibn Saud took the view that Palestine is an Arab land and that the Jews have no right in it. During 1937-40, he corresponded with the Germans for possible arms shipments to Palestine. He also helped the Palestinians financially.

On January 8, 1940, H. St. John B. Philby, an adviser to Ibn Saud, proposed a plan to the king which was aimed toward solving the Palestinian problem. His solution was that Palestine should be left to the Jews and the displaced Arabs should be resettled elsewhere at the expense of the Jews who would place a sum of £20 million at the disposal of the king for this purpose.[1] When he presented the plan to Ibn Saud, the king told Philby to keep the matter strictly confidential; according to Hafiz Wahbah, Saudi Arabia's ambassador to Great Britain, Ibn Saud told Philby to keep the plan strictly confidential because if other Arabs heard about it he (Philby) would be killed.[2]

Ibn Saud left Jeddah on February 12, 1945, on the American destroyer Quincy, for the Great Bitter Lake on the Suez Canal to meet with President Roosevelt who was returning from the Yalta Conference. He took the opportunity to tell President Roosevelt of his concern over Palestine. At the end of the conversation, Roosevelt gave Ibn Saud his assurance that he would do nothing that might prove hostile to the Arabs, and "the U.S. Government would make no change in its basic policy in Palestine without full and prior consultation with both Jews and Arabs."[3]

Ibn Saud met Churchill, after getting Roosevelt's approval on the shore of Lake Karun. Churchill told the king that his policy was to keep Palestine as a "national home for the Jews." Ibn Saud vehemently disagreed with him and told him that Palestine is an Arab country and the Jews have no right to any part of it.[4]

Violence has been endemic in the Arab-Jewish relations in Palestine since the Ottoman days. Both the Turks and the British found their task of administering Palestine was complicated by the dissensions between

Muslims and Jews. With the creation of the State of Israel the main interest of the Arabs lay not necessarily in the destruction of the new state, contrary to the accepted belief that the Arab nations fought Israel with a view to destroying it, but rather in the internal Arab competition for power and position inside the Arab world. In 1948, "This competition was mainly between Egypt, Syria, and Saudi Arabia on the one hand and the two Hashemite Kingdoms of Jordan and Iraq on the other."[5]

When the fighting broke out in May 1948, "Saudi Arabia, not having a frontier contiguous with Palestine, did not send troops or take an active part."[6] It only dispatched some token units of its troops to Palestine.[7] Saudi Arabia cooperated in the boycott of Israel but did not cut off oil production nor cancel the American oil concession.[8] King Ibn Saud became so incensed over the American recognition of Israel and the dishonoring of the pledge given him by President Roosevelt that he refused to accept a $15 million U.S. Export-Import Bank loan.[9] Ibn Saud felt that the European Jews who had suffered so much from Nazi policy of extermination should return to their home countries in Europe since the defeat of Germany in World War II had removed the Nazi threat.

In 1956, when Britain, France, and Israel attacked Egypt, "Saudi Arabia ordered general mobilization and called on all Arabs to oppose the attack. Diplomatic relations with the aggressors were severed . . ."[10] Saudi Arabia also took a strong stand in favor of Arab nationalism and supported Egypt in her nationalization of the Suez Canal.

The third war between the Arabs and the Israelis broke out in June 1967. It was strictly a war between Israel on the one side and Egypt, Syria, and Jordan on the other side. The defeat of the Arabs by Israel during the six-day war apparently did not disappoint King Faisal, due to his strained relations with Egyptian President Nasser. "But aware of Arab Nationalist sentiments, King Faisal called for the annihilation of Israel and sent troops to Jordan, although they did not engage in extensive fighting."[11]

Maintaining the already stated position of Saudi Arabia, its representative at the United Nations, Jamil Baroody, stated that "the whole creation of Israel was illegal and immoral, to say the least."[12] As to the future, he saw it written in the past: Conquerors had come and gone from Alexander the Great, the Romans, and the Crusaders, to the mandatory powers. Baroody said of them:

> Conquerors all, but the bones of their troops have, throughout the ages, bleached our soil, and their works are mere relics for the tourists to see. They are gone—all gone—those conquerors of yore. But the Arabs, in spite of all the vicissitudes that have shaped their destiny, still remain—approximately a hundred million of them.[13]

At the Khartoum meeting of the Arab heads of state, convened at the end of the 1967 Arab-Israeli war, Saudi Arabia pledged $140 million as its contribution to the Arab cause. These major cash subsidies to the

Arab states in the war and to Palestinian commando groups contributed to financial crisis in Saudi Arabia in 1969. Even then the desert kingdom fulfilled its financial commitment made to advance Arab interest against Israel.

After the immediate shock of the defeat in 1967, Nasser had assumed that what had been lost by war could be restored only by war. Nasser's successor, Sadat, also reached the conclusion that there was no way to break the stalemate other than war. In an interview with Newsweek, Sadat declared in April 1973 that "everything in this country is now being mobilized in earnest for a resumption of the battle."[14]

Like other Arab leaders, Sadat thought of the Western energy problem and the leverage which the Arab world could exercise vis-a-vis Israel in connection with the oil supply. "But unlike King Faisal of Saudi Arabia he was not willing to wait until Western dependence on Arab oil would increase and could be used more effectively as a political weapon."[15]

Following the outbreak of fighting in October 1973, Saudi forces were placed on alert. A small contingent of Saudi forces crossed into Syria, although they were never reported as fighting. However, Saudi Arabia's major role in the war was its lead in employing its oil wealth as a political weapon. On October 17, 1973, Saudi Arabia joined with ten other Arab oil-producing nations in reducing by 5 percent each month the amount of oil exported. On October 18, Saudi Arabia independently cut oil production by 10 percent to put pressure on the United States. On October 20, it announced a total halt of oil exports to the United States.

From what has been said so far, a clear pattern emerges that Saudi Arabian support in the Arab-Israeli dispute has been mostly symbolic through token forces dispatched to help Arab combatants in their fighting against Israel. Of late, however, after years in the shadows, Saudi Arabia has suddenly become an important country when one considers the importance of oil as a primary source of energy among the advanced nations. Since the beginning of the oil embargo in 1973, Saudi Arabia finds itself in the position of an oil superpower. It was quite gratifying for King Faisal to find himself treated as a statesman of great power and consequence. Above all, Saudi Arabia's newfound wealth has been a source of great financial support to Egypt—the biggest and most powerful Arab nation in terms of population, and politically and militarily speaking. It is Egypt alone that can determine the diplomatic and political terms which other Arab states may accept in the settlement of the vexed Arab-Israeli conflict. It was said of Faisal:

> For one thing, Faisal is no less a patriot than any other Arab. He feels a genuine sympathy for the displaced Palestinian people and he is an obsessive foe of Zionism. As keeper of the Holy Places of Mecca and Medina, he is especially adamant about recovering East Jerusalem, the third Holy City of

> Islam. On the other hand, Faisal does not particularly want a running quarrel with the West which, he believes, is instrumental in Saudi Arabia's plans to industrialize and become a thoroughly modern nation.[16]

Certainly the mood and activity in Saudi Arabia indicated that Faisal was earnestly interested in liberating Palestine, particularly Jerusalem, from the occupation of the Israelis. "When Faisal says, 'I want to go and pray in Jerusalem before I die,' that's the sort of thing that still counts in the Arab world."[17]

Besides being an archconservative and religious fundamentalist, King Faisal was firm in his belief that the creation of a Jewish state in the heart of the Muslim world was due to the communist-Zionist-imperialist conspiracy against Islam. King Faisal said:

> We cannot and will not ignore the forces which oppose our efforts today. There are the evil forces of imperialism and the sinister forces of Judaism and Zionism and the forces of Communism. As to imperialism, it opposes our preaching of Islam because it knows that Islam is a religion of brotherhood, a religion of peace, a religion of love, and a religion of equality. . . . As to the forces of Zionism, they know that cooperation between Moslems would put an end to the evil expansionist ideas of international Zionism in Islamic and Arab countries. . . . As for the Communists, they are attacking us because the Islamic movement is going to destroy all that Communism stands for . . .[18]

King Faisal believed that the Arabs, having rid themselves of the Ottoman and European imperialism, should not replace it with Zionism and communism. It was because of King Faisal's emphasis on imperialist-Zionist-communist conspiracy against Islam that he wanted to foster the ideal of pan-Islamism as an effective barrier against the three enemies of Islam which, in the opinion of the king, were threatening Arab and Islamic culture in the world. For the survival of the conservative monarchy in Saudi Arabia King Faisal wanted to be in the forefront of the Islamic movement and of the struggle with Israel. The new rulers of Saudi Arabia have vowed to follow the basic stand of King Faisal toward Islamic solidarity, Arab unity, and the liberation of Jerusalem. This satisfies the younger generation in the Arab world which generates pressures and revolutionary appeals to annihilate Israel.

In summary of the above discussion, one could say that Saudi Arabia, like other Arab states, has compromised partially its basic stand against Israel. The original stand of the Arabs was to dismantle Israel completely from the Middle East. This stand has been modified by all Arab nations and they now urge Israel to withdraw within 1967 borders.

SAUDI-ARAB RELATIONS

The Saudi Arabian policy in the Arab world can be characterized as dominated by two factors: pan-Arabism and anti-Zionism. Both factors are set in an Islamic cast and are similar to the policies of most other Arab states. Pan-Arabism focuses on the affinity of the Arab people and anti-Zionism is sustained in order to oppose Israel. Saudi Arabia wants not only to be recognized as leader of the Muslim world in general and the Arab world in particular but also as the guardian of the holiest Islamic places, the organizer of the pilgrimage, and the new financier. In this manner Saudi Arabia could involve all Islamic countries in a new war with Israel; however, geographical and other obstacles may prevent such a unity, except that Islamic sentiment will certainly be on the Arab side as it has been in the past.

In the 1930s soon after the formation of the modern Saudi kingdom, when recognition in the Arab world was sought, Saudi Arabia entered into treaties with neighboring Arab states on the basis of Islamic friendship and Arab brotherhood which were aimed at closer relationships with the then three independent Arab states—Yemen, Iraq, and Egypt. One such treaty, the Treaty of Taif, providing for a settlement of differences through negotiation, ended the border war with Yemen in 1934. In 1936, a treaty of Arab brotherhood and alliance was concluded with Iraq.

With reluctance, in 1945, Saudi Arabia joined the Arab League, which was formed by Britain to strengthen the relationship between member states. King Ibn Saud agreed to join the league on the explicit guarantee of each member state that the independence and sovereignty of each would be safeguarded and that each would refrain from attempting to change the form of government of any member state. Since then Saudi Arabian relations have been oriented to the individual member states rather than to the Arab League as a whole.[19]

When Israel was created in 1948 Saudi Arabia joined other Arab states in opposing the establishment of a Jewish homeland in the Arab world and made token contribution to the Arab war effort against Israel. Politically and economically the desert kingdom has consistently maintained its commitment to the principle of Arab solidarity in opposition to Israel. However, the military capability of Saudi Arabia is so limited that it is not in a position to help the Arab war cause in a big way. As will be seen below, the Saudi financial contribution to help other Arab nations fight the war against Israel has been significant.

The importance of Egypt in Saudi Arabian relations can hardly be exaggerated. The founder of modern Arabia, King Ibn Saud, made his first official tour abroad to Egypt. Economically torn by the first Arab-Israeli war and corruption, King Farouk was counting on Saudi economic assistance. Ibn Saud made a generous contribution—an annual subsidy of approximately $4 million.[20] Following Ibn Saud's death in 1953, his successor, King Saud, began to follow a more active policy in Arab affairs, making a bid for renewed identity in the Arab world. The revolutionary regime of President Nasser was recognized by Saudi Arabia soon after its formation. Another step taken by Saudi Arabia was to sign a five-year mutual defense pact with Egypt and Syria. A joint command of Saudi Arabian, Egyptian, and Syrian armies was formed. Saudi Arabia joined other Arab states in their opposition to the Baghdad Pact. King Saud became an advocate of neutralism.

Saudi Arabia was not consulted when the Suez Canal was nationalized by President Nasser in 1956. This, of course, antagonized King Saud. Yet the king contributed $1 million to help Egypt overcome its financial crisis which had developed as a result of frozen Egyptian reserves in British banks. He also broke relations with Britain and France in protest against their invasion of Egypt. It seemed that the Saudi-Egyptian friendship was not to last long. Nasser's concept of Arab socialism, born out of the simple pragmatic necessity of meeting the economic problems of Egypt, was not taken lightly by the Saudi ruler who was apprehensive of revolutionary incursion into his kingdom.

Following his visit to the United States in January 1957, King Saud turned away from Nasser and associated himself with the conservative kings of Jordan and Iraq.[21] King Saud expressed his concern over Nasser's growing popularity in the Arab world and his pro-Soviet maneuvers. President Nasser accused the Saudi king of being a reactionary. With the formation of the United Arab Republic (UAR) between Egypt and Syria, Saudi relations with the UAR worsened further. Saudi Arabia accused Nasser of interfering in the internal affairs of other Arab nations. Subsequently, a Saudi plot to assassinate Nasser was revealed.[22] While Egypt was turning more and more toward the Soviet Union, Saudi Arabia was renewing its traditional friendship with the West. Thus began a long period of confrontation between the two Arab nations.

Although the local power distribution had been more or less stable since the creation of Israel, there was sharp ideological confrontation between the conservative and revolutionary regimes. On this score Saudi-Egyptian relations deteriorated further when Egypt sent its troops and gave military aid to the new Republicans in Yemen. Saudi Arabia decided to support the Royalists. A war between the two Arab states developed in their bid to help warring Arabs. In this Arab-Arab war, Arab soldiers killed fellow Arabs. The Egyptian air force bombed Saudi Arabian towns and villages. Riyadh broke diplomatic relations with Cairo and created a joint defense council with Jordan.

Following the ouster of Saud, Faisal became the king of Saudi Arabia. The new king was more foresighted and pragmatic. He began a rapprochement between Saudi Arabia and Egypt. Diplomatic relations were restored. Faisal realized that the survival of a conservative monarchy in the midst of revolutionary fervor depended on cooperation with the largest state in the region. Arab radicalism came to be personified by Nasser. Therefore, the best chance for continued domestic stability and territorial integrity of Saudi Arabia depended on its understanding and friendship with Egypt. A period of reconciliation began. Faisal and Nasser agreed to settle the Yemeni crisis peacefully. An agreement reached in Jeddah during Nasser's visit presaged the withdrawal of Egyptian troops from Yemen. This represented a major Egyptian concession to Saudi demand.

The outcome of the 1967 Arab-Israeli war was most favorable to Israeli territorial gains. Saudi Arabia, so long considered safe from direct Israeli threat, was now fearful that the Jewish state might decide to launch an attack against the desert kingdom. On the political front the Arab heads of state, meeting in Khartoum, decided that their future strategy toward Israel should be based on three conditions known as the three "nos"—no direct negotiation, no formal peace, and no recognition. On the economic front, King Faisal proposed that the Arab oil-producing countries help to compensate for Egyptian and Jordanian war losses. The three oil-rich monarchies, Saudi Arabia, Kuwait, and Libya, agreed to put up a combined total subsidy of $392 million a year for Egypt and Jordan.[23] King Faisal's initiative in offering the handsome contribution was to obtain concession from Egypt by removing the remaining Saudi-Egyptian differences over the issue of Yemen. Thus the common threat of Israel removed one bitter struggle between Saudi Arabia and Egypt. It also ended a decade-long cold war.

After the Khartoum conference Faisal's role in the Arab-Israeli conflict changed:

> As anti-Zionist as he was anti-Communist, the King lavishly subsidized Arab governments battling Israel. He grew even more bitter against Israel in recent years, most often making no distinction between religious Jews (whom he professed to respect) and political Zionists.[24]

However, until the death of Nasser in 1970, Saudi relations with Egypt were cool but correct. Following the death of Nasser the radical doctrines of Arab socialism suffered a setback in the Middle East. Egypt, under Sadat, is fully preoccupied with a settlement of the Arab-Israeli dispute. In 1972, King Faisal visited Cairo and held wide-ranging discussions with President Sadat. Acceding to Washington's request Faisal exhorted Sadat to expel the Russians from Egypt:

> In the spring of 1972, Washington suggested to Faisal that if he would help persuade President Anwar Sadat to reduce the Russian presence in Egypt, America would press Israel to withdraw from conquered Arab territory.[25]

The strongly anti-Zionist monarch Faisal was bitter about the U.S. support for Israel but admired the Americans for their anti-Communist posture. Although the Nixon administration did not succeed in persuading Israel to withdraw from the Arab-occupied land, Sadat under pressure from Faisal expelled the Russians from Egypt and de-Nasserized the economy by dismantling much of the state ownership of property, including Saudi Arabian property nationalized by Nasser. Faisal was not willing to finance a regime that mixed communism with Islam. To qualify for Saudi assistance Egypt adopted an anti-Communist stand. Saudi Arabia, in association with other Arab oil producers, began to pay for arms purchases made by Egypt and Syria. This has been done by Saudi Arabia and other oil-rich Arab nations without bookkeeping as Arabs traditionally look after their brothers. They also offered $1 billion each to the two confrontation Arab states for financing their reconstruction programs. By using the oil embargo and by becoming the leading financier of the Arab war effort, King Faisal turned into an Arab hero almost overnight. Faisal's money and Sadat's manpower began to work hand in hand to liberate the Arab territories. Barring opinions and feelings concerning the Arab-Israeli conflict, it was American influence that spread again in the Arab world via Faisal's petrodollar and through Sadat's politics of moderation.

The Faisal-Sadat entente had other compelling reasons. They are deeply rooted in inter-Arab politics. Faisal's strongest critic in the Arab world was Libya's President Qaddafi whom Sadat considers "a hundred percent sick man." Qaddafi tries to play Nasser's role in Arab politics but without the latter's charisma; he has assumed the role of custodian of the Arab cause. By cultivating Sadat, Faisal minimized the principal appeal Qaddafi has had for a union with Egypt.[26]

Saudi relations with the other monarchies in the Arab world and the gulf states are most cordial. Before the British withdrawal from the gulf, Saudi Arabia voiced support for the idea of the United Arab Emirates. On its part, Saudi Arabia has amicably resolved its dispute over the Buraimi oasis and has waived its claims to the hinterland of Abu Dhabi. Saudi Arabia is channeling $100 million to the People's Democratic Republic of Yemen (PDRY) through Egypt.[27] The situation in PDRY is complicated as it is the main supporter of the Soviet-equipped guerrillas in the Dhofar province of Oman. Not only is Oman receiving military help from Iran to combat the rebellion in Dhofar but it has also received aid from Jordan, Pakistan, and Britain. The Sultan of Oman has offered base rights to the United States. Saudi assistance to the Dhofar rebellion is an anticlimax to its policy of anti-communism, but to counter Iranian support of Oman and being under the counterinfluence of Egypt the desert kingdom is funding PDRY to help left-leaning guerrillas in Dhofar.

The king of Saudi Arabia wants to preserve his royal image as a leader of the Arab world and the financier of the popular Arab cause. He is also desirous of guaranteeing the survival of his conservative monarchy among the multitude of extremisms in the region. To this end, Saudi Arabia has improved relations with radical regimes in Syria, Iraq, and Algeria. It has allowed the Palestine Liberation Organization (PLO) to open an office in the kingdom. By supporting PLO Saudi Arabia wants to have its influence spread into the rank and file of the organization.

PLO claims to have 90 percent support of the Palestinians who are being used as a focal point in the Arab-Israeli struggle. In accordance with the Saudi policy of Islamic brotherhood and anti-communism it does not recognize the PLO faction led by George Habash, a Christian, who is also known as a Communist.

Since the 1973 oil embargo Saudi Arabia has been playing a leading role in maintaining unity among various Arab nations. It took the initiative in ending a period of bitter feuding between Iraq and Syria on the one hand and Syria and Egypt on the other. Saudi Arabia has been busy also in ameliorating differences between King Hussein of Jordan and PLO leader Arafat. Following Secretary of State Kissinger's shuttle diplomacy technique Saudi Arabian Oil Minister Yamani has been flying from capital to capital in the Arab world to mediate boundary disputes, sharing of waters from the Euphrates River, and other quarrels. It is Saudi's moderate leadership and financial support that might prevent the spread of radicalism in the Arab world despite the existence of the Arab-Israeli conflict.

Much of the blueprint for the Arab world of the future is being drawn in Saudi Arabia and Egypt today. In spite of his conservatism, King Faisal remained a hero to the Arabs who looked to the future. This he was doing by financing PLO and by offering much needed economic assistance to Egypt's President Sadat and other Arab leaders. For the moment, Egypt and Saudi Arabia are in the ascendant. They are known as the minimalists because they are prepared to compromise with Israel and live and let live with the Jews in their midst. More important, Faisal probably joined with Sadat because if he had failed to, the maximalists—the Palestinian commandos, as well as Iraq, Syria, and Libya—might have gained ascendancy in the Arab world. It is a shrewd investment on the part of the Saudis, for otherwise dangerous revolutionary fervor is thus channeled into an exclusively anti-Israeli direction.

SAUDI-AMERICAN RELATIONS

Saudi Arabia maintains close relations with the United States. And, aside from disagreements over American support of Israel and the oil embargo of 1973, these relations have remained cordial since officially established in 1940. Saudis have been dealing with American individuals since 1933, when King Abd al-Aziz granted Standard Oil of California, renamed Aramco, an exclusive 60-year oil concession. Aramco represents the largest single American investment in any foreign country—more than $2 billion in Saudi Arabian oil production facilities since 1933.

The story of Saudi relations with the United States is one of expanding contacts through private American individuals and companies. Official contacts were minimal and developed slowly:

> The oldest steady nonofficial American contact with Arabia was through the Persian Gulf missionaries sponsored since the 1890s by the Dutch Reformed Church. From their three centers at Bahrain, Muscat and Kuwait less than two dozen missionaries exerted an influence far beyond their numbers. Not that they made many converts, for their score approached zero, but through their medical work they afforded the only association thousands of common folk and their rulers had with Americans. It is perhaps not excessive to say that they helped to prepare a better climate for the American oil men who followed later.[28]

From their bases in the Persian Gulf several of the medical missionaries established relations with King Ibn Saud. They were then permitted to visit Saudi Arabia frequently to treat the ailing king and other patients, but they were not allowed to expand their evangelistic endeavors.

The most important successful contact was made by Charles R. Crane, an American minister and a philanthropist, in the early 1930s when Saudi Arabia was severely hit by the world depression which had drastically decreased the number of pilgrims. The people of India, the East Indies, and others could no longer afford pilgrimages and naturally, this meant a decline in the Saudi government's revenue which had been five million pounds in normal years and had now dropped to two million pounds. By this time the government was in debt for 300,000 pounds, and the salaries of officials, troops, and police had fallen in arrears.[29]

It was in 1931 that Crane arrived in Saudi Arabia to discuss various economic possibilities there with the king. But the king's uppermost interest was in finding water. Crane came back to the United States and sent K. S. Twitchell to Arabia. Twitchell's report on water was pessimistic, but encouraging as regards mineral resources and oil.[30] Twitchell advised Ibn Saud to await the outcome of oil discovery in neighboring Bahrain. He then returned to the United States for the purpose of promoting capital investments in Saudi Arabia and got Standard Oil of California interested in an oil exploration project.[31]

Standard Oil obtained the oil concession on May 29, 1933, with the following payments to the Saudi government:

> (a) Before the discovery of oil in commercial quantities . . . a loan of £30,000 gold immediately; a loan of £20,000 gold after eighteen months; a rent of £5,000 gold per annum.
> (b) After the discovery of oil in commercial quantities: two loans each of £50,000 gold, at intervals of one year; a royalty of £0.2 gold per ton of oil produced; 200,000 American gallons of petrol (gasoline) and 100,000 gallons of paraffin (Kerosene) per annum.[32]

The money came at an opportune time just when Saudi Arabia was suffering from the lack of pilgrimage traffic. Standard Oil, taking a great risk, paid handsome dividends to the Saudi king.

When a new concession agreement was signed between Aramco and Saudi Arabia on May 31, 1939, the king had quite an advantageous offer from Japan, but he preferred to continue his association with the Americans as "it had the advantage of assuring the economic development of the country without incurring political liabilities."[33] The Americans had an image of disinterested humanitarianism in the mind of the Saudi king, because of their missionary, philanthropic, educational, and cultural activities in the Middle East.

Thus private American oil investments provided the starting point of contemporary Saudi-American friendly relations. Aramco had carried out vast operations without direct official backing by the U.S. government, but government policy changed following the outbreak of World War II. In 1940, Italy bombed Dhahran; pilgrim traffic came to a halt; and Saudi Arabia desperately needed financial assistance. At this critical juncture the U.S. Federal Loan Administration advised President Roosevelt that Saudi Arabia could not receive aid under the lend-lease program. The president, therefore, suggested to the British government that they continue to make financial assistance available to Saudi Arabia, since the Middle East was their sphere of influence.

The U.S. government as personified by secretaries Ickes and Knox, wanted the United States to take over the British predominance in oil concessions in the Middle East. However, it was Aramco which successfully stopped the government from following up their plans for government ownership, or part ownership, of the oil companies in Saudi Arabia. The company had as its great weapon "the traditional free enterprise suspicion of any government interference with a penetration into business."[34]

Saudi Arabia, which adopted a policy of neutrality during the war, refused to be drawn within the orbit of the Axis powers. Ibn Saud appealed to Aramco and to the British and American governments to help him out:

> He pointed out that by adopting their wartime priorities Britain and the United States had deprived him of expected oil royalties and asked for a $30,000,000 loan to be delivered in five yearly installments. His financial plight was so desperate that he threatened to cancel the concessions if he failed to obtain the required funds.[35]

It was then that an Aramco representative met with President Roosevelt in April 1941 and tried to obtain his approval for a government loan to Saudi Arabia despite the fact that the desert kingdom was not directly involved in World War II. It was decided that the United States would request Britain to make funds available to Saudi Arabia. Britain, herself, was receiving lend-lease from the United States, so that the aid to

Saudi Arabia amounted to indirect lend-lease from the United States. On February 18, 1943, two days after conferring with the petroleum administrator for war, President Roosevelt declared Saudi Arabia eligible for direct lend-lease assistance, which amounted to $17.5 million. In addition, 22.3 million ounces of silver were lend-leased which were needed to mint riyal coins.[36]

The U. S. government's financial assistance together with about $10 million in loans from Aramco[37] was several times the annual loss on pilgrimage revenues and made it possible for Saudi Arabia to avoid bankruptcy, thereby assuring the kingdom's political stability.

Meanwhile, formal diplomatic relations between Saudi Arabia and the United States were established in 1940. Although neutral, the Saudi king was the only independent ruler in the Middle East friendly to the Allies. By accepting American financial assistance Saudi Arabia not only compromised its neutrality but turned increasingly to the United States for help in modernizing its military forces and in developing its resources.

While two U.S. missions—one agricultural and the other military—were sent to assist Saudi Arabia, negotiations were being undertaken for the lease and construction of an airbase at Dhahran. Begun in 1944 and completed in 1946, the Dhahran airfield was a vital link between Cairo and Karachi, facilitating the prosecution of war against Japan. As the war progressed, Saudi Arabia was drawn closer to the Allies. After having conferred briefly with President Roosevelt on the U.S.S. Quincy in the Great Bitter Lake in Egypt, King Abd al-Aziz declared war on the Axis powers in March 1945. Subsequently, Saudi Arabia took part in the U.N. conference at San Francisco and, with the formation of the United Nations, Saudi Arabia became one of the charter members.

The first foreign trip undertaken by Ibn Saud was to meet with President Roosevelt in the Great Bitter Lake. "Afterward Ibn Saud frequently referred to Roosevelt in terms of highest praise, and the gift of a luxurious airplane no doubt deepened his friendly feelings."[38] These cordial relations have been marred by one problem—Israel. The U.N. partition resolution of 1947 on Palestine and the American pro-Israeli policy have produced tension in the Saudi-American friendship but have not caused a break in relations.

In 1946 the Export-Import Bank granted Saudi Arabia a $10 million loan. During the next year Crown Prince Saud visited the United States. President Truman presented him the order of the Legion of Merit and a citation for meritorious services to the Allied powers during World War II. In 1948 the U. S. Navy entered the Persian Gulf for the first time and paid a courtesy call on Dammam. The U.S. legation at Jeddah was raised to embassy level in 1949. By 1951, Americans had helped link Dammam and Riyadh by rail. The same year, Saudi Arabia was included in the U.S. Point Four Program of Technical Assistance. The two countries moved still more closely together by signing, on June 18, 1951, a defense agreement under which Saudi Arabia extended American usage right at the Dhahran airbase for five years in exchange for an expanded military training program which also enabled the kingdom to buy American arms under the Military Assistance Act. The defense agreement contained a provision for renewal.

Commercial production of oil began in 1945 and by 1950 Saudi Arabia emerged as the second largest producer in the Middle East. Saudi Arabia's revenue increased from a meager $300,000 in 1917 to about $90 million in 1950.[39] Anxious to develop the kingdom, Ibn Saud put special emphasis on communication, transportation, electrification, agriculture, water supply, education, and hospitals. To this end the American Mackay Corporation was entrusted with the job of building a powerful radio station in Jeddah. A Saudi Arabian air service linking the Red Sea with the Persian Gulf was established under a contract with Trans World Air Lines (TWA). With American assistance a huge pier was built in Jeddah and new harbors were developed at Dammam and Ras Tanura. The Saudi government launched an irrigation network with American help. Artificial reservoirs and water pipelines were constructed. Roads were constructed and paved between Jeddah, Mecca, and Medina with American assistance. To improve health standards the Saudi government bought surplus hospital equipment from the U.S. Army.

What happened could be described as an intensive development of economic, technical, military, and diplomatic contacts between Saudi Arabia and the United States. The rapid expansion of the oil industry by American companies did for Saudi Arabia what European armies and commerce had done for Turkey, Iran, Syria, and Egypt during the nineteenth century. Oil brought Americans to Saudi Arabia and took the Arabs to the United States in increasing numbers:

> In all of Saudi Arabia's relations with the United States Government the Arabian American Oil Company has played a major role. Company officials frequently serve as informal advisers to the King and his ministers and perform the functions of an unofficial ambassador in Washington, where the company maintains an office.[40]

However, occasionally Saudi-American friendship has been strained. The Soviet Union, which had been the first country to recognize the Wahhabi regime in 1926 and the first state to convert its consulate to a legation in 1929 with a Muslim as minister, took Saudi-American friendship as a wholesale American penetration of the kingdom. (The Soviet legation in Saudi Arabia was closed in 1938 when sharp differences arose between the Wahhabi regime and the Marxist regime of the USSR. To date no diplomatic relations have been established between the two countries.)

Relations between Saudi Arabia and the United States became further complicated after the death of King Abd al-Aziz in 1953. In 1954 a dispute over the transport of oil broke out when Saudi Arabia awarded an oil transport contract to Aristotle Onassis, the late Greek shipping magnate. Aramco claimed the use of its tankers was part of the concession; Aramco, however, won the case four years later. Saudi Arabia requested that the United States stop its technical assistance program under the Point Four agreement because the American allocation to the kingdom was small in comparison with that to Israel. The Dhahran airbase agreement expired in June 1956 and it was extended on a monthly basis.

While visiting Washington in February 1957, King Saud endorsed the Eisenhower Doctrine and agreed to a five-year renewal of the airbase agreement in exchange for continued U.S. arms aid. The United States agreed to train the 15,000-man Saudi army and to double its strength in numbers.[41]

In March 1961 Saudi Arabia informed the United States that it would not renew the airbase agreement which was to expire within one year. The chief cause of the nonrenewal, as stated by the Saudi government, was U.S. aid to Israel. On April 2, 1962, the United States relinquished its usage rights to the airbase and turned it over to Saudi authority.

Saudi-American relations suffered further decline over the Buraimi oasis dispute and the recognition of the Republican regime in Yemen. Saudi Arabia turned to the Eastern bloc for arms after the United States refused to supply it with weapons during the revolution in Yemen in 1962.[42] However, the close relations between the two countries were not completely impaired by these developments. Shortly after the Republican coup in Yemen, Faisal visited Washington and met with President Kennedy. The president explicitly declared U.S. support for the integrity of the Saudi kingdom. When the Egyptian forces bombed Najran, a Saudi town, in January 1963, the United States reaffirmed its support of the integrity of Saudi Arabia, and the Pentagon renewed its training program of the Saudi armed forces.

Saudi Arabia has put considerable store in close and friendly relations with the United States. The United States has sold Saudi Arabia military vehicles, and other equipment, and the U.S. military mission continues to provide training in the use of these equipments. The Saudis are spending more than $1 billion dollars on American weapons.[43]

The United States considers Saudi Arabia a voice of moderation in the area and for this reason wants to safeguard the integrity of the oil kingdom. The Saudis, on the other hand, have expressed their willingness to base their relations with the United States on the same freedom-loving aims. Saudi concern is being stimulated by the growing supply of Soviet arms brought into southern Yemen and Iraq. The United States is helping Saudi Arabia develop a modern airforce and navy and has initiated an equipment sale and a private training program because of this seeming threat to Saudi security. Speaking before the Subcommittee on the Near East and South Asia of the Committee on Foreign Affairs, Assistant Secretary of State Joseph J. Sisco said:

> We are going forward with the Saudis with plans to help them develop a modest naval force of small ships suitable for the

> protection of their extensive coastlines; we are selling F-5 Freedom Fighters to replace older American aircraft; we are now initiating a sales program to reorganize and equip elements of the National Guard, basically a paramilitary internal security force; and American companies are continuing to assist in improving Saudi Arabia's air defense capabilities. Recently, we have agreed in principle to consider the sale of a limited number of F-4 aircraft to Saudi Arabia.[44]

Saudi friendship with the United States and the attendant growth in its wealth had enhanced King Faisal's moderating influence and prestige in the Arab world. At the same time, by championing the cause of Palestine, King Faisal had added new dimensions to his popularity. The decisive leadership was now apparently passing from Egypt to Saudi Arabia. Not only were the Egyptians and other Arabs beginning to pay attention to the deeds and words emanating from the feudal monarch but the rest of the world was commencing to flatter him as he became the number one banker of the universe. "It has been calculated that within three years Saudi Arabia will have greater financial reserves than the United States, Western Europe, and Japan combined."[45]

The only ingredient of conflict—the liberation of Jerusalem—continues to cause bitterness among the oldest and the newest superpowers, the former being the United States—an industrial and military superpower—and the latter being Saudi Arabia—an oil and monetary superpower.

During the 1967 Arab-Israeli war, Saudi Arabia had reluctantly joined the Arab effort to withhold oil only after extensive pressure from Nasser, but lifted the embargo soon after its imposition. In 1973, it joined with other Arab oil-producing nations in imposing the oil embargo. Indeed, Saudi Arabia foreshadowed its major role in the October war when on September 4, 1973, King Faisal told the Americans via U.S. television that American support of Israel "makes it extremely difficult for us to continue to supply the United States' petroleum needs and even to maintain our friendly relations."[46]

On October 18, 1973, King Faisal's decision to reduce oil production by 10 percent was announced over Radio Riyadh.[47] Next day President Nixon asked Congress to supply Israel with $2.2 billion in military assistance. The response of the king was to impose a total embargo on shipment of oil to the United States and the Netherlands,[48] and to slash oil production by 26 percent.[49]

It is interesting to note how Aramco operated the oil embargo for Saudi Arabia. For a long time, the oil companies, considered as instruments of American foreign policy toward the oil-producing nations, "were actually used as instruments of Arab policy"[50] during the embargo. U.S. Senator Frank Church, chairman of a Senate subcommittee investigating multinational corporations, made public certain U.S. Department of State documents, hitherto classified, which revealed that the oil companies were indeed used for a long time as arms of American foreign policy.[51] But, in 1973, Aramco took a small symbolic step by cutting off oil supplies to its own homeland the moment King Faisal desired it. Obviously, Aramco carried out the instructions of King Faisal, fearing reprisal for failure to comply with the monarch's order.

On March 1974, after five months, the embargo was lifted. The king was satisfied that the United States was working seriously for Israeli withdrawal from the occupied land and Saudi Arabia returned to its preembargo production of 8.5 million barrels a day, promising to meet America's oil needs.

In return, the United States promised to help industrialize the Saudi kingdom. A set of broad economic cooperation agreements between the United States and Saudi Arabia was signed in Washington on June 8, 1974, a few days before President Nixon was scheduled to visit the kingdom. The agreement, signed by Secretary of State Kissinger and Saudi Arabia's Prince Fahd Ibn Abdul Aziz al Saud, provides for the establishment of a Joint Economic Cooperation Commission and a Joint Security Cooperation Commission.[52] As an initial step the United States is selling several squadrons of F-5E and F-5F jet fighters in a $756 million package deal to Saudi Arabia.

Saudi Arabia continued to tilt toward the United States still further, as was evidenced during President Nixon's visit to the kingdom on June 15, 1974. In an unprecedented and undiplomatic manner the Saudi king denounced Nixon's critics at home, saying that anyone who stood against the president was causing "the splintering of the world."[53] In farewell remarks before the president left Saudi Arabia, the king expressed "full confidence" that Nixon would succeed in removing all the "blemishes" standing between Arab countries and America.[54] The king said:

> And anybody who stands against you, Mr. President, in the United States of America, or outside the United States of America, or stands against us, your friends in this part of the world, obviously has one aim in mind, namely, that of causing the splintering of the world, the bringing about of mischief, which would not be conducive to tranquility and peace in the world.[55]

On leaving Jeddah for Syria President Nixon said that "our friendship . . . now develops into an active partnership."[56] As part of this new partnership program the president promised military aid to Saudi Arabia. The president said, "If Saudi Arabia is strong and secure . . . it will enhance the chances for peace."[57]

Yet, just the day before, on June 14—the first day of President Nixon's visit to Saudi Arabia—King Faisal told him bluntly that there would be no lasting peace in the Middle East while Arab lands were occupied. The king said:

> There will never be a real and lasting peace unless Jerusalem is liberated and returned to Arab sovereignty. . . . The injustice and aggression which were wrought upon the Arabs of Palestine are unprecedented in history, for not even in the darkest ages had a whole population of a country been driven out of their homes to be replaced by aliens.[58]

The king, like his predecessor King Saud, had been a friend of the United States, but he had been uncompromising in his opposition to American policy toward Israel. The successive Saudi monarchs have had special relationships with the United States, which are now being institutionalized in many fields, from defense and finance to technology and education.

Although everything is predicated on a settlement of the Arab-Israeli dispute, King Faisal told Newsweek's Arnand de Borchgrave in an exclusive interview in September 1974:

> We do not want to do anything that will hurt America. But if our new special relationship is to remain viable, the U.S. must not do anything that will hurt us and the Arab world.[59]

The other remaining source of misunderstanding between Saudi Arabia and the United States is the soaring price of oil. On this, Saudi Arabia, which initiated the memorable oil embargo, has been pressing other OPEC members hard ever since the embargo was lifted to lower the prices of oil. With the backing of the king, Saudi Oil Minister Yamani made a public commitment to abandon artificial fixing of oil prices and hold free auction. However, he was overruled by equally or more important forces within the Saudi government, as well as by the pressure of other OPEC members.

While the Ford administration's sharp warning to oil-producing nations to cut prices drew angry responses from the Iranians and Kuwaitis, the Saudi oil minister told a conference in Washington that it was time to "stop these noises nowadays and start talking seriously about the problem."[60] Indeed, in all seriousness, Minister Yamani called President Ford's "Project Interdependence" a bid "for cooperation rather than confrontation."[61]

Not content with these measures, King Faisal, in a most unusual gesture, wrote a personal letter to the Shah of Iran urging him to join Saudi Arabia in lowering the oil prices, but the Shah refused.[62] King Faisal was concerned that a worldwide depression might hurt the American and Western economy which would improve the relative position of the largely self-sufficient Communist nations. While the leaders of the non-Communist world have been struggling against the impact of the soaring prices of oil, the Soviet Union has welcomed the oil price increase.[63] Despite the professed fear of communism, the Saudis have not yet acted alone to lower the prices of oil.

Although no unilateral Saudi move to lower the prices of oil is anticipated, King Faisal told Kissinger, who had an audience with the monarch in Riyadh, that "Saudi Arabia will take up the effort to push down the price of oil in concert with other petroleum-exporting nations."[64] Acting in concert with other OPEC members, particularly with Iran, for which a series of unpublicized meetings between the representatives of the two countries were being held to work out a joint strategy,[65] Saudi Arabia was capable of bringing down the prices of oil. (In 1974 Saudi Arabia produced about 8.5 million barrel/day of oil and Iran's production was about 6.5 million barrel/day. Their combined output was about half of all OPEC production.)

Although Saudi Arabia's policy toward the liberation of Jerusalem from Israeli occupation remains unchanged, it was a significant development that for the first time King Faisal personally pledged to lower the price of oil without linking it to a territorial settlement in the Middle East.

The Saudis have maintained that it is pointless to accumulate huge sums of petrodollars if it only leads to American and the West's economic ruin. Consequently, they have begun to recycle or relend surplus funds into American treasury bonds, the stock exchange, real estate, etc., as a friendly gesture, which attitude the Saudi king adopted after a prolonged consideration of his special relationship with the United States.

Although many Americans seem to be unaware of it, King Faisal was among the few Arabs who remained basically friendly to America. He was not a radical but a conservative anti-Communist. He was not seeking the destruction of Israel, but only the return of the occupied territories, especially Jerusalem. He did not want to cut off oil supplies to the United States but was reluctantly forced to do so by other Arabs and by the action of President Nixon in announcing arms aid to Israel at the height of the Arab-Israeli war.

Cordiality between the United States and Saudi Arabia has been based on the combination of vast oil resources of the kingdom and the advanced technological skills of the Americans. Underlying this cordiality has been the endeavor of Aramco and the king to maintain mutually beneficial relations. Now as Saudi Arabia is about to acquire complete ownership of Aramco, the Saudis do not intend to expel Aramco's management, for they still need American technology. They want "Aramco to survive as a profit-making corporation that extracts and refines Saudi oil. But henceforth Saudi Arabia will be in charge."[66]

The special relationship between Saudi Arabia and the United States, commencing in the 1930s, has been maintained with freedom to follow policies not always compatible with each other's objectives. The early years of this relationship witnessed a king desperately in need of American money to maintain his hold on the kingdom. In all probability, in a normal situation, the relationship will continue to be as cordial as before, based on a new concept of mutual gratitude and sovereign equality between the military and industrial superpower and the oil and financial superpower. But an atmosphere of uncertainty hangs over Saudi-American relations as a result of the discussion in the United States of the possibility of military action against the desert kingdom in case renewed Arab-Israeli fighting leads to a reimposition of the oil embargo with disastrous results for the economies of the non-Communist industrialized states.

The postembargo special relationship between Saudi Arabia and the United States represents the highest stage of mutual cooperation. Military and technical cooperation may strengthen the interdependence even more. But there are problems. Saudi Arabia is not dependent on U.S. foreign aid, but it is purchasing American technology, expertise, and arms. The United States is buying Saudi oil. Both are able to diversify their sources of respective supplies. In such a situation both nations have retained their complete independence. The idea that once Saudi Arabia is dependent on American technology it will be difficult for it to remove itself from this dependency is not very convincing. Should the occasion arise, Saudi Arabia, a closed and singularistic society, would possibly forego its military modernization and economic development programs in order to appease its Arab neighbors. In such a case Saudi Arabia could better preserve the monarchy and maintain its security by siding with the Arab cause and by preventing dissension in Arab ranks. Meanwhile, by maintaining a cooperative relationship and by supplying much-needed and much-desired technology and arms the United States can hope to gain a firmer foothold politically and economically with Saudi Arabia.

NOTES

1. H. St. John B. Philby, Arabian Jubilee (New York: John Day, 1953), pp. 212-13.
2. Hafiz Wahbah, Khamson Am Fi Jazerat al Arab (Cairo, 1960), p. 179.
3. William Eddy, F.D.R. Meets Ibn Saud (New York: American Friends of the Middle East, 1954), p. 36.
4. Wahbah, op. cit., p. 159.
5. Yair Evron, The Middle East: Nations, Superpowers and Wars (New York: Praeger, 1973), p. 16.
6. Sydney N. Fisher, The Middle East (New York: Alfred A. Knopf, 1959), p. 583.
7. Fred J. Khouri, The Arab-Israeli Dilemma (Syracuse: Syracuse University Press, 1968), p. 70.
8. Fisher, op. cit., p. 534.
9. Ibid., p. 538.
10. Ibid., p. 539.

11. Congressional Quarterly, The Middle East: U.S. Policy Israel, Oil and the Arabs (Washington, D.C.: Congressional Quarterly, April 1974), p. 66.

12. United Nations, General Assembly, Doc. A/PV. 1555, (New York: July 17, 1967).

13. Ibid.

14. Newsweek, April 9, 1973.

15. Walter Laqueur, Confrontation: The Middle East and World Politics (New York: Bantam Books, 1974), pp. 46-47.

16. Newsweek, February 18, 1974.

17. Ibid.

18. Ministry of Information, Kingdom of Saudi Arabia, Faisal Speaks, Book 12, n.d., pp. 48-49.

19. George A. Lipsky, Saudi Arabia: Its People, Its Society, Its Culture (New Haven: HRAF [Human Relations Area Files] Press, 1969), p. 141.

20. Norman C. Walpole et al., Area Handbook for Saudi Arabia (Washington, D.C.: American University, 1971), p. 164.

21. Charles D. Cremeans, The Arabs and the World: Nasser's Arab Nationalist Policy (New York: Frederick A Praeger, 1963), p. 109.

22. Ibid.

23. Malcolm H. Kerr, The Arab Cold War (London: Oxford University Press, 1971), p. 139.

24. Time, April 7, 1975.

25. New York Times Magazine, March 24, 1974, p. 52.

26. Nadav Safran, "The War and The Future of the Arab-Israeli Conflict," Foreign Affairs, January 1974, p. 221.

27. Jim Hoagland, "Shah's Force Jolted in Oman: Iranian Troops Test Prowess in Rebel War," Washington Post, December 16, 1974.

28. John A. De Novo, American Interests and Policies in the Middle East, 1900-1939 (Minneapolis: University of Minnesota Press, 1963), p. 355.

29. H. St. John B. Philby, Arabian Days (London: Robert Hale, 1948), pp. 289-90.

30. Benjamin Shwadran, The Middle East, Oil and the Great Powers (New York: Frederick A. Praeger, 1955), p. 288.

31. K. S. Twitchell, Saudi Arabia (Princeton: Princeton University Press, 1958), p. 222.

32. George Kirk, Survey of International Affairs: The Middle East in the War 1939-1946 (London: Oxford University Press, 1952), p. 355.

33. George Lenczowski, The Middle East in World Affairs (Ithaca, N.Y.: Cornell University Press, 1962), p. 549.

34. Shwadran, op. cit., pp. 339-40.

35. Ibid., pp. 550-51.

36. Raymond F. Mikesell et al., Foreign Investment in the Petroleum and Mineral Industries (Baltimore: Johns Hopkins University Press, 1971), p. 107.

37. Ibid., p. 78.
38. Lenczowski, op. cit., p. 553.
39. Ibid., p. 555.
40. Mikesell et al., op. cit., p. 81.
41. See for more information Dale R. Tahtinen, *The Arab-Israeli Military Balance Today* (Washington, D.C.: American Enterprise Institute for Public Policy Research, 1973), p. 21; Saudi Arabia in 1972-73 had an army strength of 36,000.
42. *Washington Post*, September 11, 1962.
43. Dana Adams Schmidt, *Armageddon in the Middle East* (New York: John Day Co., 1974), p. 96.
44. *New Perspectives on the Persian Gulf: Hearings before the Subcommittee on the Near East and South Asia of the Committee on Foreign Affairs, House of Representatives, Ninety Third Congress* (Washington, D.C.: U.S. Government Printing Office, 1973), p. 3.
45. Schmidt, op. cit., pp. 93-94.
46. *Congressional Quarterly*, op. cit., p. 66.
47. Allan T. Demaree "Aramco Is A Lesson in the Management of Chaos," *Fortune*, February 1974, p. 58.
48. Ibid., pp. 58-59.
49. Ray Vicker, *The Kingdom of Oil* (New York: Charles Scribner's Sons, 1974), p. 104.
50. *Washington Post*, August 25, 1974.
51. Ibid.
52. *Washington Post*, June 9, 1974.
53. *Washington Post*, June 16, 1974.
54. Ibid.
55. Ibid.
56. Ibid.
57. Ibid.
58. *Washington Post*, June 15, 1974.
59. *Newsweek*, September 30, 1974.
60. *Washington Post*, October 5, 1974.
61. *Newsweek*, October 7, 1974.
62. *Time*, October 14, 1974.
63. New York *Times*, October 13, 1974.
64. *Washington Post*, October 14, 1974.
65. Ibid.
66. Edward R. F. Sheehan, "Mastermind of Mideast Oil," *Reader's Digest*, September 1974, p. 84.

CHAPTER 5

THE ROLE OF OIL IN SAUDI FOREIGN RELATIONS

The fact that the oil resources were developed by American companies may have influenced the general orientation of Saudi Arabia in the East-West struggle. While Saudi Arabia, as all other Arab nations, claims to be neutral, its destiny undoubtedly lies more with the West than with the Soviet bloc. For one thing, it is the West and Japan which provides the markets for Arabian oil and not the USSR, which is presently an oil surplus country. Saudi Arabia has no diplomatic relations with any Communist state.

The huge oil reserves of Saudi Arabia have made its relationship with its neighbors and the non-Communist nations of great significance. Saudi Arabia is oriented toward the West for purposes of defense, industrialization, and the sale of oil. Previously, it had been in general conflict with Britain over the claims of oil in her protectorates, which Saudi Arabia also wanted. But with the beginning of the 1973 Arab-Israeli war the British government followed a neutral policy and Saudi-British relations improved perceptibly.

Basically, since the discovery of oil deposits, Saudi Arabia's foreign policy has changed little, but there have been important developments. The basic Saudi Arabian policy objectives are to prevent the encroachment of communism and radicalism into the Arabian Peninsula, to defend general Arab interests, to promote solidarity among Muslim nations, and to maintain cooperative relations with other oil-exporting and oil-consuming nations.

Foreign policy is determined by a small group of men, headed by the king himself. An example of this was the oil embargo and production cutbacks of 1973:

> When Arab oil ministers met in Kuwait during the October war to consider the oil embargo and production cutbacks, Saudi Minister Sheikh Zaki Yamani diligently inserted several discreet loopholes in the final decision to give the King more

> flexibility. But the King, infuriated by the subsequent U. S. announcement of $2.2 billion in military aid to Israel, rejected the escape clauses and ordered a stunned Yamani to cut production three times as deeply as the joint Arab decisions required.[1]

In the field of foreign policy, King Faisal was determined to avoid any collision with Western materialism and lead his country to affluence while insulating it from socialism and communism as well as from liberal democracy. Saudi Arabia, which by 1980 may possess $200 billion in foreign investments and currency reserves, is emerging as a major source of world financiers.[2] Eventually, the Arabs, led by Saudi Arabia financially, "will join America, Russia, China, Western Europe, and Japan in the fellowship of global powers."[3] And, when it comes true, it will be a direct contribution of the massive oil resource that Saudi Arabia possesses.

SAUDI PETROLISM

Taking into consideration this background on the general foreign policy orientation and formulation related mostly to oil in Saudi Arabia, let us study in more detail Saudi petrolism within OPEC, and within the Arab and the international systems.

Petrolism—whether viewed as supplementary to Arab military incapability or as a pressure on the consuming nations for the survival of energy-hungry modern civilization or as a tool for escalating the price of oil, or as a combination of these—inspired King Faisal and other Arab leaders to make use of it. He used this valuable instrument to maximize politicomilitary expediency and enhance economic advantage by exploiting a highly vulnerable situation in the oil-consuming industrialized world. Of all the political forces working in the contemporary Arab world—Arabism, anti-Zionism, nationalism, radicalism, populism—petrolism shapes the oil policy which affects not only the destiny of the Arab but also that of the whole world.

The Arab petrolism has not been so much an instrument of pressure as the coming to fore of the weakness of the oil-consuming nations, particularly the United States. This is clearly illustrated by the fact that the Saudi Arabians waited until 1973 to use their oil weapon against the United States, at a time when American dependency on imported oil had reached more than one-third of supplies. Even then the timing of the October war and the use of oil as a weapon of diplomacy was not optimal from the point of view of gaining maximum benefit. Had the Arabs waited until 1980, when U.S. dependency on Arab oil under the existing consumption rate would have reached its peak, petrolism would have been more effective in the strangulation of the industrial world.

Despite pressure from Nasser and other radical Arabs, King Faisal had waited long, but perhaps not long enough, to use his oil weapon against the United States. However, being encouraged rather prematurely by the first Arab victory at the initial stage of the war against Israel, he decided not to delay his decision any further, since the direction of the war was being quickly reversed by the subsequent Israeli success in occupying Arab territory in the heart of Egypt.

Recognizing the vulnerability created by the dependence of industrial consuming nations on the Arab oil supply, Saudi Arabia decided to employ the oil weapon to compensate for the weakness of Arab military capability. The oil embargo was carried out in such a way as to maximize political interests and financial gains. The political goal was accomplished through the concerted efforts of OAPEC and the economic objective was geared more closely with the interests of Arabs and non-Arab oil exporters.

The Saudi policy of petrolism to maximize political and economic gains through oil began in 1959 when an Arab Oil Congress was formed. The viewpoint of the congress in its second meeting in 1960 was in sharp conflict with the oil companies. The formation of an inter-Arab body to coordinate the oil policy was hampered further by the factiousness in the Arab body politic. Although an inter-Arab body of oil-producing countries could not be formed at this stage, an organization (OPEC) of all oil-producing developing nations was set up largely by the initiative of Venezuelan Perez Alfonso and the then Saudi oil minister, Abdullah Tariki, a petrocrat trained at the University of Texas. Since then OPEC has emerged as a power center, bent toward the correction of past injustices and the control of the oil resources.

OPEC has had a universal perspective and its formation had the effect of dissociating Arab oil policy from larger Arab political and economic interests. Since an Arab oil organization could not be formed, Saudi Arabia began to work within OPEC to serve the cause of the universal interests of all oil exporters, to the exclusion of particular Arab interests. Soon Tariki was ousted and Saudi policy, both inside and outside OPEC, supported the policies of the oil companies to some extent.

In the mid-1960s Iraq demanded a greater control of the Iraq Petroleum Company. Saudi's policy was to prevent such control and to work in cooperation with the oil companies. As expressed by Oil Minister Yamani:

> I believe that we in Saudi Arabia have set an example worthy of emulation as regards the establishment of a truly fruitful relationship with the oil industry, sustained through cooperation and negotiation. I should like to differentiate between the right to take unilateral action and the actual exercise of this right. There is no need for Saudi Arabia, for the time being, to think of taking unilateral action. It is against our

general philosophy of doing business and inconsistent with the friendly atmosphere which characterizes our relations with the oil companies at present. And I am sure that the oil companies operating in Saudi Arabia have no interest whatsoever in shaking our faith in this philosophy by showing us that other means are more rewarding in safeguarding our oil interests.[4]

In accordance with its philosophy of free enterprise, as an alternative to nationalization, Saudi Arabia called for participation agreements with oil companies. Participation became an official policy of OPEC, despite the fact that Syria, Algeria, Iraq, and Libya had begun to nationalize their petroleum resources. Being anti-Communist Saudi Arabia is an ardent advocate of private ownership of property. Even then it took the view that "the producers had been made to feel that they were foreign to their own natural resources, with neither control over them nor say in how they were utilized."[5] The oil companies ignored OPEC and for a decade the companies maintained a stable underpriced oil. This was intolerable for the oil-exporting countries, since prices of commodities imported by them from oil-consuming countries were steadily rising. It would seem that the oil-exporting countries were double losers: selling at less competitive prices and buying manufactured goods at inflationary prices. Instead of bargaining for an oil price increase, Saudi Arabia began to work out the participation formula which, when implemented fully, would assure more revenues for the desert kingdom. While Saudi policy wanted some moderation in the rank and file of OPEC, the latter considered it an attempt to sabotage a unified stand on oil nationalization. However, without Saudi cooperation OPEC could not adopt a more nationalistic policy.

Presenting his favorite participation proposal at the Middle East Institute's 26th Annual Conference in Washington, Yamani said:

> Oil is the chief source of energy today, and so long as there are sufficient reserves of it to keep it flowing, it will continue to be so. . . . Whenever the energy shortage is the subject of discussion, the kingdom of Saudi Arabia, as the owner of the largest proved oil reserves in the whole world and of vast unexploited areas where the geological and geophysical tests indicate the presence of huge amounts of oil, emerges as an important factor.[6]

Yamani's interest in participation rather than nationalization stems basically from the fact that Saudi Arabia alone cannot handle the vast oil industry at the present time. In the future when the Saudi national oil company acquires the necessary technology involved in the oil industry, the participation proposal may turn out to be an outright nationalization. Of course there are other factors which have deterred nationalization of the oil industry in Saudi Arabia. The country has a personalistic monarchy which is primarily interested in oil revenues. The king seems to be dependent on the United States and other Western countries for maintaining his power. However, the participation formula would allow the national oil company to "become a connecting factor between producers and consumers and play exactly the same role played at present by the international oil companies."[7] Petromin acts as the agency for participation.

During the six-day Arab-Israeli War, Iraq initiated an oil boycott which was also joined by Saudi Arabia. But King Faisal was opposed to a prolonged boycott and within a week wanted the oil companies to renew exports, when the war ended in six days. However, he did not agree to the Iraqi proposal to withdraw funds from American banks and nationalize Aramco. Faisal refused to attend the Arab summit meeting at Khartoum unless Nasser agreed to settle the Yemeni dispute and stopped calling the king a reactionary. On both the counts Nasser conceded. At the Khartoum conference Faisal agreed to give financial assistance to Egypt and Jordan when Nasser gave up his demands concerning oil boycotts and withdrawal of Saudi funds from American banks.[8] Thus ended a long period of hostility between a reactionary king and a progressive president, thanks to the Saudi Arabian petrodollar and the Israeli threat.

Tariki's policy was to seek a greater voice in control of oil resources and an economic nationalism undermining the monopolistic control of the oil industry by foreign companies. However, his ouster nipped the seed of nationalism in the bud. Yamani's conservative policy and cooperation with oil companies thwarted progress in the direction of national control. Increasingly he was coming under pressure from other OPEC nations to take full advantage of the changing market situation favoring a sellers' control of oil.

In 1967 the oil-producing Arab countries on their behalf also formed a group, OAPEC, which without competing with OPEC, is in a position to defend more particularly the Arab point of view; that is, its action may sometimes be on a regional political scale not found in OPEC. OAPEC includes Saudi Arabia, Kuwait, Algeria, Libya, Iraq, the United Arab Emirates (UAE), Bahrain, Egypt, Syria, and Qatar. (The members of OPEC are Venezuela, Iraq, Iran, Saudi Arabia, Kuwait, Qatar, Abu Dhabi, Algeria, Nigeria, Libya, Indonesia, Ecuador, and Gabon [associate member].) OAPEC contemplates various forms of action in such fields as industrial development, pipelines and tanker transport, production control, and marketing procedures.

Yamani, who was one of the founders of OAPEC, defines the purposes of the organization as follows:

> (1) To keep oil activity within the Organization with a view to protecting the member states from precipitate decision and making oil a genuine weapon to serve the interests of the producing countries and the Arab countries in general.
> (2) To create opportunities for joint investment in the oil resources of the member states.

> (3) To build an economic bridge between the Organization and the consuming countries so as to create an outstanding economic structure which we can utilize to expand our markets and thereby establish a stronger political center of gravity for the states of the region.[9]

This gives us an idea of the perception of the role that OAPEC can play and the political weight expected to center in the hands of its members. The status quo with the oil companies is still maintained by Saudi Arabia to secure the continuation of exploration, drilling, and marketing. Gradual changes in the relationship will parallel the rate of improvement in the operations of the Saudi national oil company.

The growing testimony in the United States by oil company lobbyists of its increasing dependence on Saudi Arabian oil and the lingering shadow of the Arab-Israeli conflict forced Saudi Arabia to adopt a policy favoring the OPEC decision to increase prices of oil on the one hand and the OAPEC policy of using oil as a political lever in the Arab-Israeli conflict on the other.

Politics per se and the politics of oil became closely intermingled and inseparable in the inter-Arab and Arab-Israeli conflicts. Added to this was the magnitude of the American support of the Israeli war efforts, which forced Saudi Arabia to assume control of the oil companies and implement production cutbacks and embargo supplies through them.

It has been suggested that, with the huge revenues that it is acquiring as a result of increased prices, Saudi Arabia has three alternatives for disposing of its income: (1) to float the money in the floating currency markets, (2) to recycle the funds in the consuming nations which are unable to pay for higher oil prices, and (3) to invest in the consuming countries to help offset their balance-of-payment problems. Saudi Arabia, in addition to following the three alternatives, is exploring a fourth alternative, that is, developing its own or Islamic financial institutions. The establishment of the Islamic Development Bank in Jeddah in August 1974 is an example. Saudi freedom of action in this regard is controlled somewhat by its inability to run business organizations alone, owing to its lack of managerial and technical know-how.

In terms of financial reserves Saudi Arabia is the richest nation in the world, but in terms of per capita income it is among the poorest. Saudi Arabia, like other oil exporters, is suspicious of transferring its monetary surpluses to the industrialized countries which could mean replacing the oil concession regime by a new form of exploitation of Arab oil resources. Because of a lack of coherent policy of investment in the developing nations most Saudi and other Arab surplus oil revenues are invested in the Western world.[10] As part of the Third World, Saudi Arabia hates exploitation and, much more, to be called an exploiter. In dealing with the oil companies Saudi Arabia does not like to be viewed as taking advantage of the consuming countries' pressing need for its oil when asking for higher prices.

Even though there are special and long-term ties between Saudi Arabia and the United States, "Saudi Arabia too became disgusted with American policy in the Middle East and took under serious consideration the imposition of restraints on its oil production."[11] In the summer of 1973 it was apparent that Saudi Arabia was willing to use its oil as a weapon to bring pressure on the United States to moderate its policy in the Middle East. Yamani first conveyed the Saudi decision to the American government during a visit to Washington in April 1973.[12] King Faisal himself issued a warning that Saudi Arabia would find it "difficult to continue cooperation with the United States in the petroleum field unless Washington moves toward a more balanced policy in the Middle East."[13]

The Arab oil-producing states were under immense pressure from other Arab countries, particularly Egypt, to use their petrolism against Israel as the tide of the October war turned against them. During several days of debate preceding the oil embargo the Arab oil-producing nations decided to cut their oil production by 5 percent and to cut another 5 percent each month until Israel withdrew from the occupied land. "At the same time, all deliveries of oil to the United States, Canada, and the Netherlands, three nations deemed to constitute a citadel of Zionist support, were immediately embargoed."[14] At another meeting a further production cut of 25 percent was announced. The Common Market countries were exempted from the effects of further cuts as they favored settlement of the Arab-Israeli conflict on the basis of U.N. Resolution 242.

To administer the oil embargo Saudi Arabia divided the countries of the world into groups of friendly and unfriendly nations and the rest. The friendly group included Britain, France, and Spain which were to continue to receive oil at the preembargo level. The unfriendly nations such as the United States and the Netherlands were to receive no oil at all. The rest of the world was to receive whatever oil was left. Aramco was given the responsibility of implementing the embargo against its own home government which the company did faithfully.

The Arab countries in the autumn of 1973 were producing 19.1 million barrels of oil a day. The United States, which consumed 17.2 million barrels a day during this same period, imported 1.8 million barrels a day, or less than 10 percent of its total consumption, from the Arab countries. This included direct and indirect shipments from the Arab world.[15] Situations in Europe and Japan were worse. Western Europe received 65 percent of its oil from Arab countries; Japan was getting 50 percent of its oil needs from the Arab world.[16]

Following the embargo, a 400 percent escalation in the price of oil is vastly increasing the revenues of the major producers of the Middle East. In 1974 alone, the oil revenues of eight Middle Eastern and North African countries—Saudi Arabia, Iran, Kuwait, Libya, Iraq, Abu Dhabi, Algeria, and Qatar—approached \$72 billion as against \$17 billion in 1973. Of the \$72 billion the Saudi share is \$21 billion which is, of course, the largest.[17] Even though these nations want to maximize development of their economies and get the most for their oil, the strategies they are adopting seem to differ according to their perceptions of the energy situation in the world.

SAUDI ARABIA AND THE ENERGY CRISIS

King Faisal had foreseen the energy crisis and had tried to persuade the United States government to head it off by sending emissaries to Washington to convince American policy makers of the merits of an evenhanded policy in the Mideast but without any success. The main Saudi plea as reported by Dana Adams Schmidt was this:

> Saudi Arabia now has ample income and capital for all its needs and development. It does not need, for its own purposes, to expand production of oil any more. Indeed, the inevitable rise in the price of oil makes it advantageous for us to leave the oil in the ground as long as possible. You, the United States, are urging us, however, to expand production of oil rapidly to meet your needs. We are now under great and growing pressure not to accede to your demands, and even to buy back our production, unless you are willing to adopt a more evenhanded policy in the Middle East.[18]

With the precipitation of the energy crisis, Saudi Arabia's voice on matters of oil supply and pricing has become powerful. King Faisal's opposition to the increase in prices and his instructions to Yamani to bring them down were firm.[19] The king, however, appreciated the need to maintain the unity of OPEC whose solidarity brought so much benefit to him and other oil producers. However, OPEC is restrained by Saudi Arabia, which has the largest proven reserves of oil in the world and is a frequent dissenter to proposals to increase prices.

The Saudi attempt in 1974 to lower prices was motivated by the realization that the oil price level—a 400 percent increase—was too high for the world to bear. Such an increase was creating monetary instability and inflation of a kind that was against the interests of Saudi Arabia and other oil producers. King Faisal, a fierce anti-Communist, was equally concerned that the free enterprise economy of the West should not be seriously weakened by an energy crisis while the basically energy-surplus Communist countries, particularly the Soviet Union, maintained a stable economy. The king's preoccupation with communism is basic to Saudi foreign policy and accounts for its close relationship with the West. In addition, he was well aware that the Western nations could best provide his kingdom with the technology needed for Saudi development and defense. Paradoxically, technology and equipment were the basic tools used by the colonial powers to subjugate the Arab world.

Like other oil-producing nations Saudi Arabia sees its oil deposits as finite with conceivable exhaustion in the future. At the same time, there is the possibility that development of new forms of energy may mean that oil production, now voluntarily foregone, will never be made up.

The trend in Saudi policy in 1975 was to maintain existing price of oil by curtailing supply. Saudi Arabia has reduced its production of 8.5 million barrels a day to 6.5 million barrels, due to the drop in demand of the world market. Reducing oil production offers one of the rare opportunities, like the embargo, in which Saudi Arabia can please both the religious leadership that emphasizes the return of Jerusalem to the Arabs and the political radicals simultaneously. It also responds to a third very influential group in Arab politics, the technocrats, who believe in the logic of increasing revenues through lower production and higher prices. This was precisely the theme of Yamani's television appearance in Washington in April 1975.[20]

Saudi Arabia believes that oil producers and oil consumers are linked by a common interest and that oil producers can provide the energy-hungry industrialized nations with oil in return for the purchase of arms and goods available in the industrial world. What is sought here is an oil-for-arms-and-technology exchange. Saudi Arabia also looks to the West for safe and productive uses of its vast supply of oil money.

The Saudis are adopting the sophisticated concept of "linkage" in their dealing with the West. They are determined to be treated as the monetary and oil superpower they are fast becoming. The Arabs believe that the West "must adjust to the shift in financial power, and learn to deal with us as equals,"[21] and, therefore, are determined to link oil and monetary cooperation to other issues of importance to them.

To implement the decision to invest their petrodollars in American industries, a private Arab corporation with the backing of Saudi Arabia, Egypt, and Abu Dhabi has been set up for the purpose.[22] This Arab multinational corporation intends to gain control of a number of American companies in much the same way American multinationals took over many European, Afro-Asian, and Latin American firms in the past. Aware that the U.S. Congress may pass legislation to block such Arab takeover bids, a Saudi leader told Newsweek, "we will have to conclude there is nothing very special about the special relationship you have proclaimed with Saudi Arabia."[23]

Saudi Arabia knows that it cannot benefit from economic chaos as a result of the energy crisis in the world. As Yamani says, "If there's a depression, we'll never be safe—no matter how rich we are. . . . We cannot industrialize if there is a sick economy in the world."[24] To maintain and foster interdependence Saudi Arabia has been advocating a rhetoric of cooperation rather than confrontation. Saudi leaders are urging the West to enter into serious dialogue with the oil producers to work out mutually beneficial solutions to the growing economic ills of the world. Saudi Arabia has pledged substantial contribution to the

World Food Council to create a new international fund to promote production, marketing, and distribution of food, in the developing countries.

Nevertheless, there is still a gap in perception and interests between the Arab oil producers and the West. Although Arab oil producers are beginning to invest their revenues in the West, they are doing so for their own reasons and not in response to the warnings from Western politicians and economists that economic collapse looms on the globe. While the West argues over the need for recycling petrobillions, Arab policymakers view this idea with suspicion and generally oppose the concept of recycling. (The concept of recycling involves lending of surplus oil revenue by the producers to international agencies for relending to nations with balance-of-payments deficits.) Apparently the Arabs doubt the proclaimed economic doom of the West.

For Saudi Arabia the new situation is not a threat but a new economic reality—a transfer of wealth from the industrial world to a group of oil-producing nations. However, it has established a development fund of $2.8 billion for loans to developing nations.[25] Saudi Arabia along with Kuwait and Abu Dhabi have pledged more than $400 million to the Islamic Development Bank.[26] They have also declared non-Arab developing states eligible for low-interest loans. These nations are careful that loans from them are linked to specific projects, not spent only for fuel bills, as recycling means.

Many of its loans and grants to other nations are not publicized by Saudi Arabia, since there would otherwise be more and more requests from developing countries for such assistance:

> Political motives, tinged with an implicit appreciation of revenge, play a role in the Arab refusal to cast themselves as the economic cavalry galloping in to save beleaguered Third World and Western economies.[27]

Like the public investments, the private savings being accumulated by ruling families in Saudi Arabia and elsewhere in the Persian Gulf are not made known:

> The ruling families and their friends get healthy slices of the oil revenue and are increasingly putting this money into investment abroad, as well as their traditional well-funded private bank accounts in Beirut and Switzerland.[28]

Behind this basically economic discussion there are some political issues at stake: the competition between Saudi Arabia and Iran in the Middle East; and the influence and policies of the United States in the region. Having committed himself to the cause of an Arab-Israeli settlement, King Faisal was anxious not to muddy the issue with the United States on the question of the price of oil. It is also a matter of concern to Saudi Arabia that the Shah of Iran is becoming too powerful financially on the back of high oil prices.

Because of the fact that Saudi Arabia has virtually more room for expansion of oil production than the rest of the Middle East put together and the fact that high oil prices will, in the long run, encourage alternatives to oil to appear in the market, Saudi Arabia was encouraged to work for lowering the high prices of oil. How far Saudi Arabia could go to bring down the prices, either by acting alone or in concert with others, is open to question. Certainly Saudi Arabia cannot isolate itself from the storms of public opinion and criticism that might be aroused at home and in other oil-producing nations if it decided to act unilaterally to lower oil prices.

Even though the Saudis now have a fairly clear idea of what they want to do, they are not certain how to do it. Yamani, however, foresaw the problem a year before the oil embargo was imposed:

> We are now faced with two problems of unequal importance. The less important problem is that of determining the new oil prices in 1975. The other and far more serious problem is that of the expected shortage of energy supplies in the eighties. . . . It is a fact that the trend of oil prices is going upwards and there is no way of changing it. However, the OPEC countries must make only reasonable demands based on justice. The consumer countries on the other hand must accept an equitable increase in prices and absorb the whole of such increase, so as to leave the oil companies enough profits to enable them to continue their exploration activities and development of producing fields.[29]

Yamani never uses the term "crisis" in discussing the energy situation in the world. He prefers to call it an energy shortage or scarce energy. He believes, "When the age of scarce energy dawns"[30] the Saudi national oil companies will play an important role in alleviating the energy shortage in the world. In the past the oil-consuming nations have guided the policies of the producers. The 1973 oil embargo proved that under the present international climate "the producing nations have wrested control of their resources and will use them in the determination of their own destinies."[31]

INFLATION AND EXPLOITATION

The oil-exporting nations blame inflation and past colonial exploitation as the main reasons for higher price of oil. To a certain extent inflation in the Western world may be blamed for the loss in purchasing power of the petrodollar. But it is not the external inflation alone which causes genuine worries in the oil-exporting countries, since there are severe domestic inflations which are uncontrollable as more and more money is pumped into hitherto underdeveloped economies. External inflation is probably contributing to decreased savings of petrodollars in numbered and checking accounts in foreign banks. The external inflation cannot be a good rationale to raise oil prices, since the money saved this way for the future of the oil-exporting nations is at the expense of many other nations on the verge of collapse from paying extremely high petrobills.

It was the Western oil companies that developed the oil resources of the Third World. Had there been no Western technology and money available for the development of the oil resources probably the oil-exporting nations of today would not have known that oil as a primary source of energy existed under their soil. Of these nations only Saudi Arabia has a most reasonable attitude on the twin issues of inflation and colonial exploitation. The oil kingdom seldom, if ever, has branded the oil companies or for that matter the Western nations for inflation and colonial exploitation.

Saudi Arabia depends almost wholly on its one-crop economy—oil—for its prosperity. Currently, it cannot use its petrodollars for lack of absorptive capacity; therefore, it is banking on the future. In the hope of blunting the fast development of alternative energy in the United States, Saudi Arabia may want to cooperate with it, not only by selling oil at a fixed price on a long-term basis, but also to diversify its economy from the increasingly vulnerable oil resource to other sources of revenue.[32]

A base has been established for accelerating economic and social development in Saudi Arabia. The kingdom is potentially the largest holder of financial surplus in the world. On this score, Saudi Arabia has no substitute for cooperation with the West and Japan. Money not needed now is invested abroad, mostly in Britain and the United States, and in financing development projects and military imports of Egypt, Syria, and Jordan and by doling out to PLO and other nonrich Arabs and mostly non-Arab Muslim nations for their needs. Huge sums of money are wasted by the paying of allowances to the princes and rival tribes to maintain the royal family's hold over various factions and in counterbalancing the army with the tribal national guard.

Saudi Arabia and the Middle East as a whole has a dominant position in today's world oil industry. By 1985 experts contend that oil production outside the Middle East could increase by 23 million barrels a day. These new resources, it is estimated, will be almost sufficient to meet the world's demand of oil at the present projected level of declining consumption.[33] Dispersion of this additional supply from sources outside the Arab world would hamper OAPEC nations in organizing oil embargoes in the future. Besides, the high probability of the development of alternative energy sources would reduce the chances of Arab oil embargoes to nil by 1985.

Because of its successful policy of petrolism, Saudi Arabia emerged from its former subservient status in world affairs to being a major force in world politics. As a founding member of OPEC Saudi Arabia assumed the leading role in the world of petrolism by virtue of the fact that it possesses one-fourth of the free world's oil reserves, giving it self-confidence. It effectively turned Aramco, formerly an instrument of U.S. foreign policy, into an Arab instrument of petrolism.

Such petrolism helped to consolidate and strengthen its role and prestige within the Arab system. By financing the radical elements, Saudi Arabia successfully neutralized their influence in inter-Arab politics. As a result of the higher price of oil, huge amounts of petrodollars have been transferred from the oil-consuming world to the owners of oil. In this Saudi Arabia has been the greatest beneficiary and it can be assumed that there would be unwillingness to give up its newfound financial position by reducing the price of oil.

Saudi Arabia, which had long remained an insular nation, was slowly coming out of its isolation by following a progressive policy of supporting the OPEC and OAPEC causes in furtherance of their political and economic momentum. Knowing full well that too great a use of oil as an instrument of petrolism could antagonize the United States, Saudi Arabia began to plead for moderation soon after the embargo was imposed and withdrew it before any real damage was done to its existing relationship with the United States. At the same time Saudi Arabia was put in the position that it could not stand firm and not use its oil weapon against the United States for fear of losing its prestige and position in the Arab world. Therefore, Saudi Arabia followed a moderate course of petrolism.

As Saudi Arabia has become involved in international politics on a global basis, it may not be too long before the oil kingdom opens its door to the Communist group of nations. The Saudi hatred of communism is based, of course, on its incompatibility with Islam. But considering the fact that many other Arabs and Muslim nations have diplomatic relations with the Communist bloc, Saudi Arabia, to promote its independent policy in an effort to balance its international relationships, may opt for "friendship toward all and malice toward none," an American phrase, but an oft-repeated maxim of the nonaligned nations.

NOTES

1. Washington Post, May 13, 1974.
2. New York Times Magazine, March 24, 1974, p. 14.
3. Ibid.
4. Middle East Economic Survey, November 18, 1966.
5. Shaykh Ahmad Zaki Yamani, "Prospects for Cooperation Between Oil Producers, Marketers and Consumers: The Issue of Participation and After," in World Energy Demands and The Middle East (Papers delivered at the 26th Annual Conference of the Middle East Institute, Washington, D.C., September 29-30, 1972), p. 96.
6. Ibid., p. 95.

7. Ibid., p. 98.

8. Middle East Record, 1967, p. 263.

9. Middle East Economic Survey, September 13, 1968.

10. See Middle East Economic Digest, July 19, 1974. Available information on the current methods utilized by the Arab countries for their oil revenues indicates that most of these revenues are invested in the industrialized countries, particularly in Britain and the United States. Statistics for the first six months of 1974 show that $1 million on the average are invested abroad every week and the estimate shows that in 1974 these investments will approximate $45 million. Britain is one of the main beneficiaries of Arab investments and it is believed that if the trend recorded during the first half of 1974 is maintained, these investments would by the year's end reach $10 billion. A large part of the investments is made in the form of bank deposits. The rest is distributed among real estate purchases and the acquisition of stocks and various participation.

11. Tareq Y. Ismael, The Middle East in World Politics: A Study in Contemporary International Relations (Syracuse: Syracuse University Press, 1974), p. 236.

12. Ibid.

13. Christian Science Monitor, July 6, 1973.

14. Dana Adams Schmidt, Armageddon in the Middle East (New York: John Day Co., 1974), p. 212.

15. Ibid., p. 213.

16. Ibid.

17. Newsweek, July 1, 1974.

18. Schmidt, op. cit., p. 216.

19. Washington Post, October 14, 1974.

20. See the transcript of "Face the Nation," CBS Television Network (Washington, D.C., April 20, 1975).

21. Newsweek, October 7, 1974.

22. Ibid.

23. Ibid.

24. Ibid.

25. Washington Post, October 30, 1974.

26. Ibid.

27. Ibid.

28. Ibid.

29. Yamani, op. cit., p. 99.

30. Ibid.

31. Ismael, op. cit., p. 239.

32. See the transcript of "Face the Nation," op. cit.

33. See Edward R. Fried "World Market Trends and Bargaining Leverage," in Joseph A. Yager and Eleanor B. Steinberg, Energy and U.S. Foreign Policy (Cambridge, Mass.: Ballinger Publishing Co., 1974), pp. 256-58.

CHAPTER

6

OIL AS AN INSTRUMENT OF ARAB AND SAUDI POLICY

> There are these days ever-increasing worries about the not too distant future when the world is expected to face an energy shortage. Such a shortage would not only harm the world's economic progress, but would probably have adverse effects on all aspects of its civilization as well.[1]

This was the way Yamani perceived the impending energy crisis in September 1972, a little over a year before the kingdom took the lead in imposing the oil embargo. He was aware of the coming oil shortage and its resultant consequences. In 1972, plans were underway to increase Saudi oil production to 20 million barrels per day by 1980 to meet the growing energy needs of the world.[2] Saudi Arabia felt a moral obligation to provide the world, particularly the United States, with oil.[3] The benefit of mutual cooperation with the United States and the advantage of a free enterprise economy over antagonism and regimentation were foremost in the Saudi Arabian government's mind.

This moderation in Saudi Arabian oil policy was due to the lack of unity in foreign and economic policies of the Arab states. The external need for their resource was not sufficiently exploited politically until the beginning of the oil embargo in 1973. Till that time, despite the difficulty of separating economics from politics, the Arabs had tried to keep the two apart as much as possible, for fear of a possible economic loss.

The oil-exporting Arab countries, which had practically nothing to do with the development of the oil industry except that the hidden treasure happened to be in their territories, have created the greatest impact on the international oil industry today. As a group these OAPEC nations are a very important source of the world's oil supplies and they are the most militant, which poses a serious threat to the flow of oil to other nations. Without oil these countries would be of significantly less importance on the world scene. With oil, but individually, these nations have no real impact on the world stage.

First, none of the oil-exporting countries at present has a dominant position in the international oil market, although Saudi Arabia's oil resource is almost inexhaustible. It is because of this huge oil reserve that Saudi Arabia carries maximum weight in OAPEC. Not all OAPEC nations can follow supply-demand ratio in production over a long period of time due to their greater need of oil revenue. Only Saudi Arabia, Kuwait, and Libya are capable of increasing or lowering production to keep supply on a reasonable balance with demand. Added to this is the huge accumulation of petrodollars by Saudi Arabia which enables it to hold the key to OAPEC or even OPEC's success or failure. Secondly, the driving force for change is oil, which is not only shaping the structure and character of the society in the Arab world but also radically changing the economic and political significance of the region to the rest of the world. As far as can be foreseen, Arab oil is indispensable in meeting world energy demands for the first few years of the current decade (1975-85) and hence the accompanying importance of that region to the outside world.

ARAB OIL AS A WEAPON BEFORE 1973

The concept of using oil as a diplomatic weapon for the Arab nations against the West and Israel is as old as the Arab-Israeli conflict itself. Anticipating Western support of the Zionist cause, the Arab League passed a set of resolutions in June 1946, one of which called for denial of oil to the West.[4] When fighting broke out in May 1948 in Palestine, the decision to embargo the supply of Arab oil to the West was not implemented, largely due to the opposition by Saudi Arabia, "which believed that a commercial oil operation should be divorced from political consideration."[5] Considering the fact that at that time all Arab oil was being produced by the international oil companies, the Arab League realized that its member states had no voice in deciding the level of production, price, and export of the commodity.

Nevertheless, this premature attempt to use the oil instrument as a diplomatic weapon in the settlement of the Arab-Israeli conflict had some successes as noted by Lenczowski:

> . . . in a gesture of defiance toward Israel and out of solidarity with other Arab states Iraq stopped the movement of oil by pipeline to the Israeli-held Haifa terminal and caused construction of the parallel line between Kirkuk and Haifa to cease . . . and . . . boycott measures against Israel by the Arab League gradually affected the transactions of a number of oil companies with Israel.[6]

Of the four Arab-Israeli wars so far, oil figured significantly in all but that of 1948. In 1956 when Britain and France attacked Egypt, Israel joined in, making it the second Arab-Israeli war. The Arab countries not only blocked the flow of oil to Israel but also to the Mediterranean via the Suez Canal, and the pipelines from Iraq and Saudi Arabia were closed. Some pipelines were blown up by nationalist elements in retaliation for the tripartite attack on Egypt. The situation arising from the closure of the Suez Canal and the embargo on shipment of Arab oil to Britain and France produced serious consequences for their economy. Oil embargoes increased prices in Britain and France, as the oil companies had to obtain the commodity from the United States, which shipped their Middle Eastern oil products via the Cape of Good Hope which again entailed more costs. As a result oil rationing was introduced in Western Europe.

During the Suez War, in a strange duet the United States found itself in "alliance" with the Soviet Union. They played a major role in calling upon Britain, France, and Israel to cease fire and to withdraw their forces. Directly or indirectly the two superpowers urged Egypt and other Arab countries to abandon the embargo on the shipment of oil to their adversaries. Consequently, Arab attempts to use their control over oil as a diplomatic weapon for a long period of time failed, again without achieving the desired goal of settling the Arab-Israeli conflict. The oil embargo which lasted for six months had no lasting impact on the Western European economy.

In 1967, when the third round of the Arab-Israeli war began, Mideast oil became even more important to most of the industrialized world due to the fact that oil has almost replaced the use of coal since 1965. As the war broke out Egyptian President Nasser announced to the world that Britain and the United States had joined Israel in its attack upon the Arabs.[7] He also said that the U.S. Sixth Fleet helped in the Israeli attack on Egyptian airports and military bases. These accusations were not true, but at that time the Arabs took them seriously. Notices were issued to the oil companies to cease exporting oil to the countries blacklisted by the Arabs such as Britain and the United States. Germany was added to the list for its sale of gas masks to Israel. For the first time the Arab oil-producing nations shut down all their oil production. In a further retaliatory mood in Saudi Arabia and Kuwait, "bands of saboteurs assembled with explosives, ready to destroy the big companies' installations once and for all."[8] Before the oil fields could be blown up by the saboteurs Arab troops moved in to occupy them. Soldiers were ordered to stop the flow of oil, but their presence prevented sabotage. The pattern of the 1956 oil embargo was repeated by closing the Suez Canal once again, and this time it was closed indefinitely.

The European economy was not as badly hurt as in 1956, because oil was imported from Libya, Algeria, and Venezuela. The United States also exported oil to Britain and other European countries to help them meet the crisis.

The oil embargo was of short duration, for only two weeks later oil flowed again. There were several reasons for the failure of the embargo. First, because of the inaccurate propaganda, the Arabs decided to use their oil weapon as the only alternative to inaction. Second, within a few days all of them found that none of the countries had the financial strength to carry on the oil embargo without getting money from the oil companies. "Saudi Arabia was the first to feel the pinch acutely. . . . King Faisal was informed by his finance minister that there was no more money in the till, and that for once Aramco was unable to help."[9] Finally, the Arab summit conference, held in Khartoum shortly after the war, decided that oil should be used "positively" as a political weapon, the implication being that their approach was based negatively on false propaganda and other wrong notions. In any event it is worth referring to a statement in this connection made by the under secretary of the Kuwait Ministry of Foreign Affairs, who said:

> Kuwait has always subscribed to the notion that oil should be used as a weapon in the confrontation with Israel, and has, however, applied this notion ever since it established the Kuwaiti Fund for Arab Economic Development, which strengthens the Arab economy by financing development projects that have helped and continue to help Arab steadfastness . . . it is not in anyone's interest to use oil negatively.[10]

The use of oil as a weapon in 1967 had been badly handled. "Injudiciously used, the oil weapon loses much if not all of its importance and effectiveness," said Yamani.[11] On the Arab maneuvering of oil, he said further that "if we do not use it properly, we are behaving like someone who fired a bullet into the air, missing the enemy and allowed it to rebound on himself."[12]

The oil shots missed the targets because, first, the United States was not hurt by them. On the contrary, the embargo enabled the international oil companies to make handsome profits by sizeable production of U.S., Venezuelan, and North African oil. Second, despite the closure of the Suez Canal the oil companies were able to supply oil to their customers in Europe without much difficulty. Third, no quota ceilings were imposed by the Arab governments which encouraged overlifting of oil from Mediterranean and other ports.

"Oil is a formidable weapon if it is used properly," said King Hussein, the ruler of the oilless Kingdom of Jordan.[13] The proper and positive use of oil as a weapon of diplomacy can be made if sufficient funds are available to sustain the long and arduous task of implementation of an embargo and to help the have-nots in the rank and file of the Arab world.

Following the June war of 1967 the whole atmosphere of inter-Arab politics changed. Although the Arabs were not motivated by international considerations, they were beginning to give serious thought to their pan-Arab sense. The message of military defeat reached equally to the belligerent and nonbelligerent. "There could hardly be a competition for prestige when there was no prestige remaining."[14] Self-interest and rivalry were not as strong a factor deterring concrete action aimed at unity and sharing wealth as before. The Arab oil countries have been making financial contributions to the oilless and oil-short Arabs since the 1967 war.

The Arab countries involved directly in the conflict with Israel are Egypt, Syria, and Jordan. These are not oil-producing nations, although a little oil activity is seen here and there in the countries, with the exception of Jordan. These countries call on the big oil-producing Arab nations to use their oil as a weapon in the struggle against Israel in the name of Arab solidarity. For example, "in 1967 Saudi Arabia cut the flow of oil involuntarily, under pressure by Nasser, and therefore did not enforce the measure strictly and cancelled it as soon as possible."[15] However, pressure or no pressure, Arab nations in general respond to the call of Arab unity and put their oil weapon into use against Israel or the West whenever the occasion demands.

As pointed out previously, the world's oil industry structure changed in the beginning of the 1970s with the disappearance of surplus production in the United States and the growing control of multinational oil companies by OPEC nations. The United States and other industrialized non-Communist countries of the world were dependent on the supply of oil from the Arab world. Oil industries in the United States stressed the significance of Arab oil, especially that of Saudi Arabia, as a most important source of supply to the Western Hemisphere.

In 1972, the Economic Council of the Arab League undertook a study of the strategic use of Arab economic power vis-a-vis Israel and the energy-consuming world. The report did not call for an embargo but noted that restrictive oil production would bring pressure on the consuming nations to alter their uncompromising policy toward Israel.[16] Gradually Saudi Arabia and other Arab nations were coming under pressure to reduce their oil production.

Upon the advice of King Faisal, Sadat expelled the Russians from Egypt so that the United States would use pressure on Israel to withdraw back to the 1967 borders. Clearly the Nixon administration failed to fulfill the commitments it had made to Faisal. As a final attempt at a peaceful settlement, Sadat sent his national security adviser, Hafez Ismail, to Washington to prevail upon President Nixon to make good his commitment to King Faisal. It was at this time that Israeli Premier Golda Meir was visiting Washington and President Nixon yielded to Israeli pressure by making public his intention of supplying Israel with Phantom jets. This infuriated both Sadat and Faisal.

THE 1973 OIL EMBARGO

The Saudi monarch, who had gained considerable prestige and respect in the Arab and Muslim world, was very much embarrassed by the renewed American desire to supply sophisticated arms and equipment to Israel, which caused him to lose face with Sadat and other Arab leaders. Therefore, without further delay he dispatched Yamani to Washington in April 1973 where he linked oil and politics for the first time. Yamani told American officials that it was impossible for Saudi Arabia to work against the interests of its Arab neighbors. American officials did not take Yamani seriously, ignoring him in the belief that he was speaking for himself and not for King Faisal. To remove these misgivings, Faisal told Aramco's president in Saudi Arabia that he was "not able to stand alone much longer"[17] in the Arab world where pressure was accelerating for the use of oil as a weapon. The author has learned from an Aramco official in Washington that, following King Faisal's warning, the company launched a campaign in government circles in Washington, urging that the United States follow an even-handed policy in the Middle East, but without any success. The Aramco official cited the powerful Jewish lobby in the United States as an effective barrier to the company's success to change U.S. government policy of informal partiality toward Israel.

On May 15, 1973, an anniversary of Israel's creation, Algeria, Kuwait, Iraq, and Libya stopped oil production for a short time as a symbolic message to the world that Arab oil producers could and would use their oil weapon effectively against oil-consuming nations at opportune moments. About the same time Libya, taking action against foreign oil holdings, nationalized the American Bunker Hunt and took over the majority interests in Occidental and Oasis. Still, Saudi attitude showed that nothing had really changed in its policy of moderation and traditional friendship with the United States. Saudi Arabia's policy was one of wait-and-see.

The oil-producing Arab nations, such as Saudi Arabia which was the leading country in the imposition of the embargo in 1973, had already increased oil revenues and oil production beyond the current need of social and economic development. The Saudi kingdom was prepared, it seems, with ideas for imposing the embargo selectively even before the war broke out in October. On September 4, the Christian Science Monitor carried an article, by staff correspondent John K. Cooley, regarding an interview with King Faisal and his son Prince Saud al-Faisal which had been published in a Lebanese paper on the use of oil as a weapon. The Cooley article noted in part:

> . . . oil is not an artillery shell, but an enormous weapon. All economic weapons need study and time for their effectiveness to appear. Talk of using the oil weapon . . . makes it sound as if we were threatening the whole world, while it is understood that our purpose is to bring pressure to bear on America . . . but America would be the last to get hurt, because the U.S. will not depend on Arab oil before the end of the 1970s, whereas Japan and Western Europe depend on it

> now. What benefits are there . . . from arousing the fears of the Europeans and Japanese at a time when they are showing greater sympathy for us? . . . Arab policy is called upon today to persuade the American and European citizen that his interests are with the rights of the Arabs and that we do not intend to harm him, but that it is the policy of his government that is creating the confrontation (with Israel). We must tell the American and European people that we want to defend ourselves, not harm them.[18]

When the war broke out on October 6, it looked at first as though the Arabs would not need to use the oil weapons. As the war turned against the Arabs the oil ministers of the OAPEC held a meeting in Kuwait on October 17 to consider the role of oil in the Arab struggle to liberate their lands occupied by Israel in 1967. Obviously in their attempt to bail out Egypt and Syria, which were directly engaged in the war against Israel, the Arab oil ministers,

> considering that the ultimate goal of the current struggle is the liberation of the Arab territories occupied by Israel in the 1967 war, and the restoration of the legitimate rights of the Palestinian people in accordance with United Nations' resolutions,
>
> considering that the United States is the principal and foremost source of Israeli power that enabled it to continue occupying their territories,
>
> considering that the industrial nations have a responsibility of implementing the United Nations resolutions, and
>
> considering that the economic situation of many Arab oil producing countries does not justify raising oil production, although they are willing to make an increase to meet the demand in those industrial nations that are committed to cooperation in the task of liberating occupied territories,
>
> decided that each Arab oil exporting country immediately cut its oil production by a rate not less than 5% from the September production level, and further increase of 5% from each of the following months, until such a time as the international community compels Israel to relinquish occupied Arab lands, and to levels that will not undermine their economies or their national Arab obligations. [For full text of the Arab oil ministers' resolutions see the Appendix.]

October 17 was a red-letter day in the history of the world's energy crisis when all Arab oil countries joined in the war with Israel by imposing an immediate 5 percent cut in oil production and by raising that cut monthly by 5 percent until such time as Israel withdraws from the occupied Arab territories. The countries joining in the embargo were Saudi

Arabia, Kuwait, Iraq, Libya, Algeria, Abu Dhabi, Qatar, Bahrain, Egypt, and Syria. The next day, Saudi Arabia independently cut production by 10 percent. The following day Libya announced that it would raise the price of oil by 28 percent more, although there had been several small increases in the price before the October war began. Iraq announced a 70 percent increase in the price of oil. When the oil ministers met again in Kuwait on November 4, they decided that the initial reduction in output would be 25 percent of the September level and that a further cut of 5 percent would be made during each of the following months to "levels that will not undermine their economies or their national Arab obligations." Meanwhile, following massive deliveries of U.S. arms to Israel, a total halt of oil exports to the United States and the Netherlands was announced by Saudi Arabia, Abu Dhabi, Kuwait, and Algeria "and the actual diminution in supplies to other industrialized nations varied between 5 & 10 percent."[19]

While regretting the move, the Arab oil ministers said that Israel had contributed to a 5 percent reduction in the flow of oil to the world by bombing the oil terminal in Syria and forcing a 50 percent reduction of the flow of oil through the Trans-Arabian Pipeline. The ministers appealed to the world community as a whole and the American people in particular to "help us in our struggle" against occupation. Although the Arab countries wished to cooperate with all, and the oil producers were ready to supply the world with its oil needs, the time had come for condemnation of Israel's aggression, said the oil ministers.

The Arab nations' willingness to use their lubricated resource as a subtle weapon of diplomacy has been attributed to their enhanced concern about Israel, and to their additional latitude gained from the accumulated revenue:

> As long as Israel had been hemmed in within the pre-1967 boundaries, surrounded by a ring of Arab states that contained it and threatened to roll it back, countries like Saudi Arabia, Kuwait, and Iraq could feel completely safe from any Israeli threat. Whatever contribution they made to the Arab cause against Israel was made purely on the grounds of pan-Arab considerations. But as Israel overwhelmingly . . . demonstrated a . . . capacity to hurt them in a significant way . . . Their concern with Israel . . . became an investment in their own security.[20]

Following the death of Nasser and the subsequent improvement in Saudi-Egyptian relations, King Faisal demonstrated his increased ability to guide his country's foreign policy more independently:

> . . . in the past . . . Faisal feared that once he had sprung the "oil weapon," others, particularly Nasser, might be able to arrogate to themselves the right to decide when and how it was to be used.[21]

When the Arab oil-producing countries administered the embargo decision reached in Kuwait in their own individual way, Saudi Arabia separated the countries of the world designated for supply of oil into three categories,[22] which were (1) the friendly countries, (2) the unfriendly countries, and (3) all others. The friendly countries included Britain, France, Spain, Pakistan, Malaysia, and all the non-oil-producing Arab nations. The unfriendly countries included the United States, the Netherlands, Canada, Portugal, South Africa, and Rhodesia. The unfriendly countries were to receive no oil at all. The friendly countries were to continue to receive oil at the preembargo level. "When they had been served, said King Faisal, then whatever was left might be distributed to the rest of the world."[23]

As the Arab embargo went into effect a panic-like situation ensued in Europe and Japan. However, the situation was not one of absolute panic in the United States, but serious concern was noticeable during the embargo days. Both radical and conservative Arab states embargoed oil against the United States, the Netherlands, and, of course, Israel. Saudi Arabia, the most unradical of all, wanted to put pressure on the United States to minimize its support for Israel. Iraq, one of the most radical, believed that Europe, barring few countries, and Japan, which were increasingly showing a friendly attitude toward the Arab cause, should be supplied with oil. The Arabs knew that nothing spectacular could be expected from the United States, which had a declared pro-Israeli bias for obvious domestic political reasons.

With Europe and Japan gravely threatened by their heavy dependence on the Arab oil, they were subjected to serious Arab diplomatic pressure or "blackmail" as news media reports indicated. At times the Arabs issued ultimatums and oral threats that a rigorous oil embargo would be imposed on Europe and Japan if their policies toward Israel were not changed. Indeed, under the pressure, Britain and West Germany, both known for their long pro-Israeli biases, banned arms shipments to the combatants, including Israel. London even stopped American transport planes from landing on British territory and Bonn protested the shipment of American arms from Germany. Thus a rift was developing in NATO, and the United States was left alone in its support of Israel.[24] The European governments denied that they had been threatened in any way by the Arabs. As the British foreign secretary said, "There has been a great deal of talk recently about submission to Arab blackmail. . . . The Arabs have made no demand on us and we have offered no price."[25] The fact is that while Britain and other European countries were adopting a neutral position toward the Arab-Israeli conflict, the supply of oil from the Arab world was becoming progressively more assured. As the Sunday Times (London) pointed out, "Britain's oil is safe—if we behave ourselves."[26] Another important power in Europe, France, because of its special relationship with the Arabs, expected better deals from them and indeed received an almost uninterrupted flow of oil. Meanwhile, Japan attempted to retain its neutral position by continuing to support U.N.

Security Council Resolution 242 which called for Israeli withdrawal from occupied Arab lands.

How is it possible that the few desert Arab nations could seek wide compliance in their policy from Europe and Japan and bring such pressure to bear on the United States that the industrialized world suddenly moved from a wasteful habit of living toward conservation of resources by introducing gas rationing, banning weekend driving, closing service stations, lowering thermostats, and reducing driving speeds? (The lowering of thermostats and the reducing of driving speed were introduced in the United States, while the other measures were in wide practice in Europe. To save energy Britain even legislated a three-day week for work.) To answer this crucial question let us resort to what may be called a technical description of the consumption of oil in the industrialized countries of the West and Japan.

In the fall of 1973 the Arab nations were producing 19.1 million barrels of oil a day. The United States then consumed 17.2 million barrels per day and depended on export from the Arab countries for about 10 percent of it. Europe consumed 15 million barrels per day and depended on Arab sources for 65 percent of it. Japan consumed 5.2 million barrels per day, of which 50 percent came from the Arab countries.[27]

World dependency on Arab oil increased from 34 percent in 1957 to 54 percent in 1973.[28] Let us now review the impact on prices of this heavy world dependency, as OAPEC put their petrodiplomacy into use. The price increase occurred in three steps during the embargo period. Prior to the outbreak of the Yom Kippur war (or Ramadan war) on October 6, 1973, the Saudi government revenue per barrel of oil was about $1.90; when the embargo was lifted on March 18, 1974, the revenue had risen to $9.25.[29]

The first price increase, on October 16, 1973 was relatively uninfluenced by the embargo since the decision had been made prior to the embargo. The six largest oil-producing countries on the Persian Gulf—Iran, Iraq, Saudi Arabia, Kuwait, Abu Dhabi, and Qatar—announced a price increase from $3.01 a barrel on October 1 to $5.12 on October 16. The next price increase by these nations came on December 23, effective as of New Year's Day, when the price per barrel of oil was raised to $11.65. But the actual market price rose to $15-to-$17 per barrel.[30] Libya and Nigeria set a new tax reference price, ranging from $14.60 to $18.75 a barrel for crude oil.

Speaking for the oil producers, the Shah of Iran, arguing that the posted prices of oil should be adjusted upward to reflect the market price (posted price or tax reference price is, by practice, about 40 percent higher than market price), wanted "to junk the current pricing system devised by the international oil companies and base future prices on the costs of supplying alternate sources of energy."[31] Libya and Algeria wanted to push prices as high as possible. In this price escalation Iran, Libya, and Algeria were opposed by Saudi Arabia which proposed a smaller increase "because it wanted to keep the politically motivated embargo separate from OPEC's pricing policy to avoid the impression that the embargo had been imposed for monetary reasons."[32] Saudi Arabia felt that a sharper increase in price would debase the political message the Arabs had hoped to communicate to the world.

The next increase occurred in January 1974 when Kuwait announced its 60 percent participation formula and the price of oil went up by more than $1 per barrel. These increases in price are rational, said the oil producers, mainly on grounds of worldwide inflation. Inflation did not occur suddenly with the beginning of the Ramadan war; it had started long before. It appears, therefore, that taking undue advantage of the war and the embargo, Iran, in association with other leading oil producers of the region, spearheaded the move to quadruple the oil prices. As one researcher in the petroleum industry notes, "In the absence of the October War and the resulting Arab oil embargo, prices would not have risen so rapidly in so short a period."[33]

The price of oil is dubious and confusing. Behind this is the system of oil pricing which has developed over the decades. At the base is the posted price, which is a point of reference for tax and royalty collection and the buy-back price realization. The royalty is a fixed percentage of the posted price which the oil companies pay to the host countries for extraction of oil. After the royalty has been paid and the cost of production, which is about 12 cents a barrel, is deducted from the posted price, taxes are realized. The buy-back price was introduced after the oil embargo when some of the producing countries acquired 60 percent of the ownership of the oil companies. The buy-back price is set at a portion of the posted price.[34]

Oftentimes the oil-producing countries decide to increase taxes, royalties, and buy-back prices, claiming that they are not raising the prices of oil. The companies which are making exorbitant profits from oil sales are not willing to reduce their margin of profits. Thus the increase is ultimately passed on to the consumers.

In this situation some readjustments have been made, notably by Saudi Arabia, the United Arab Emirates, and Qatar. They reduced their posted prices by 40 cents a barrel but raised the tax, royalty, and buy-back prices slightly, making them almost equal to 40 cents. As a result there was no impact on the oil market and consumers continue to pay higher prices just as before.

By the time the first year of the energy crisis for the non-Communist world was nearing an end in 1973, the Arabs launched a vigorous diplomatic and publicity campaign all over the Western world and Japan. The Saudi minister for petroleum, Yamani, accompanied by the Algerian minister for power, Abdessalam, toured Western Europe, the United States, and Japan to explain the Arab oil policy. They held press conferences and met with target groups all over those countries to clarify the Arab stand on the use of oil in their struggle for the liberation of Israeli-occupied Arab territories.

Not content with this, in full-page advertisements in the *Washington Post* and the New York *Times* on the eve of the new year, Saudi Minister of State for Foreign Affairs Omar Sakkaf wished the American people, against whom the embargo was primarily directed, a "Happy New Year," on behalf of the Arab people. Appreciating the fact that the American "holiday season may have been marred by the hardships of the energy crisis,"[35] the Saudi minister brought home to the Americans the plight of the Arabs by saying, "ours is haunted by the threat of death and continued aggression."[36]

Referring directly to the Arab oil embargo the Saudi Arabian minister said:

> We cut oil supplies to the United States after the United States, which had repeatedly assured us of our rights to our lands, made massive arms deliveries to the Israelis to help them remain in our lands. We did so not to impose a change in U.S. policy in the Middle East but to demand the implementation of U.S. policy in the Middle East, as it has been repeatedly defined. We did so not to "blackmail" the American people, but to put our case to them as effectively as we knew how.[37]

Almost simultaneously, King Faisal in his first public speech since the October war called on all the Muslims of the world to mobilize their resources "to rescue our sacred places in Jerusalem from the Zionist and Communist menaces."[38] Speaking to a group of high-ranking pilgrims the king, who was the official protector of the holy places of Islam, told them that he had a special responsibility for liberating Jerusalem, which includes the mosque of Omar, Islam's third holiest—Mecca and Medina being the first and the second, respectively.

The worldwide Arab psychological warfare to justify their use of oil as a weapon in the Middle East conflict only proved how vulnerable the Arabs themselves really were. The use of oil as a weapon was creating an anti-Arab backlash that could be helpful to Israel. There was much public sympathy for Israel in many Western European countries but, in general, government policy there called for refraining from any actions that might antagonize the Arabs, lest the oil supply to Europe be endangered. Considering the neutral attitude of the Western European governments and Japan, the Arab oil states decided to end their monthly production cuts and ease restrictions on oil exports to those nations.[39] In an editorial the New York Times said:

> The Arabs have brought their oil weapon into play in the Mideast war, but cautiously and in a manner that leaves the way open for diplomatic cooperation . . . it is evident from the conduct of the Arab diplomatic delegation . . . that the leading oil states are more interested in accommodation than in confrontation with the West . . .[40]

Saudi Arabia, taking the lead in the imposition of the oil embargo, began to plead for moderation soon thereafter, and it was under the moderating influence of its Oil Minister Yamani that the embargo could be lifted on March 18, 1974, by a decision of most of the OAPEC states meeting in Vienna. He also "steered OPEC towards a freezing of crude oil posted prices for another three months . . ."[41] Following the lifting of the embargo, oil supplies to the United States were resumed. Yamani accomplished a three-month price freeze by threatening to break the united oil price front of OPEC by posting a separate lower price. To maintain the unity of OPEC—a most successful international cartel—the members agreed to Yamani's proposal to freeze oil prices.[42]

There were two OPEC meetings in the summer of 1974: one in Quito, Ecuador, in June and the other in Vienna in September. In these meetings Yamani tried to stave off price increases by threatening to increase production. The other producers countered by pleading to cut back production by an equal amount. Saudi Arabia felt that the price of oil was already high and any further increase would attract substitution of oil which would make oil reserves useless. Other producers were in disagreement with Saudi Arabia and the result was a compromise, a slight increase in price.

The notion that Saudi Arabia, the largest exporter of oil, could unilaterally bring down the prices of oil is somewhat misleading:

> An increase in export capacity by 2.5-3.5 million b/d which is all the country would be physically capable of in the next 3 years could be largely offset by corresponding reductions in countries with surplus oil revenues.[43]

Nevertheless, temporarily, the oil embargo proved to be the strongest of diplomatic weapons. "It prodded the U.S. which in turn prodded Israel, and the result was the disengagement of Israeli and Egyptian forces along the Suez Canal."[44] Although the Soviet Union, the principal arms supplier to the Arabs and their political booster, called on the Arab oil producers to maintain the embargo,[45] the Egyptians and the Saudi Arabians were eager to reward U.S. Secretary of State Kissinger for his step-by-step shuttle diplomacy in the mediation efforts to bring about a cease-fire, and subsequent withdrawal of Israeli troops from the west bank of the Suez Canal.

Most people of the world were puzzled by the Arabs' decision to use oil, a primary commodity, as a weapon in their struggle against Israel, an act which on the surface looked unfriendly to everyone. Apparently oil was used as a political weapon and not so much for economic expediency, although some OAPEC and some OPEC nations took economic advantage of it initially.

THE ARAB STRATEGY

Soon after the lifting of the embargo the Arabs expounded at the U.N. General Assembly special session in April 1974 on their oil policy objectives. Inaugurating the session, Algerian President Boumediene, who convened it, developed a "work program in five points," which was the Algerian viewpoint on the role of oil in the economic development of the oil-producing countries. The five-point program which basically included the nationalization of oil and other resources in the developing countries was announced by Boumediene after the prices of wheat and fertilizers which had doubled between June 1972 and September 1973 were reconsidered. Reacting to the Western criticism of the oil price hike, Boumediene rebutted:

> . . . the fundamental difference that explains the greatly dissimilar reactions caused by rises in fertilizer and wheat prices on the one hand, and in the price of oil, on the other, resides in the fact that the proceeds of the increase went to developed countries in the first case and to developing countries in the second . . .[46]

Addressing the special session of the General Assembly, Saudi Oil Minister Yamani declared that the oil-exporting countries would reject any attempt aimed at imposing a tutelage on oil prices on them. He added that Saudi Arabia could afford to cut its oil production by half to raise the price further, but did not do so because of its "sense of responsibility towards the rest of the world."[47] In exchange for this goodwill he wanted the industrialized world to help set up diversified industrial structures in Saudi Arabia as assured sources of income when oil runs out.

The Arab strategy, since the imposition of the embargo, has been to tie in America's relations with its Atlantic partners which are the major consumers of Arab oil. While the United States was firm in its Mideast policy the Europeans were not in a position to offer effective resistance to Arab pressure simply because of their greater dependence on Arab oil. The Arab pressure on Europe worked as a reversal of the earlier European colonial situation. Although no permanent dent was created in America's relations with its Atlantic partners, the tendency in Europe "was not to blame the Arabs, or even Israel, but to say that it was United States' Middle Eastern policy that was causing Europe to freeze this winter."[48] Indirectly, the Europeans were putting pressure on the United States to adopt an even-handed policy in the Mideast. It is through continued dialogue and closer relations with the Western European nations that the Arabs attempt to weaken Israeli ties with Europe and, through it, with the United States, the major benefactor of Israel. If the Arabs succeed in this strategy, they would win and Israel would lose a contest which otherwise has had no resolution through wars.

Another Arab strategy has been to isolate Israel from the main currents of world politics and create a sense of insecurity for the Jewish state. The need for peace is greater for Israel than for the Arabs, both economically and politically, mainly because of three new developments after the 1973 war. Because of fear of insecurity immigration to Israel is slowing and emigration from the Jewish state is accelerating. Added to this is a 25 percent inflation of Israeli currency due to the higher cost of military spending, which resulted in the devaluation of the Israeli pound by 43 percent in 1974 and again by 2 percent in 1975. The latter provides for monthly 2 percent devaluations if the economy continues to decline. At the same time foreign investment in Israel slumped by about 60 percent.[49] Fears of an Arab boycott are seen as a factor inhibiting foreign investments in Israel. Politically, the Arabs were effective in isolating Israel internationally. Following the October war many Afro-Asian nations which had earlier established diplomatic links with Israel severed those relations. Thus, Israel is becoming insecure, while the Arabs have in their possession tremendous newfound wealth in the form of petrodollars. As the Arab-Palestinians continue to gain more sympathy in the world there is consequently a reduction in sympathy toward Israel. The Arabs were successful in cutting off support for Israel by the United Nations Educational, Scientific, and Cultural Organization in 1974. In the future Arabs might initiate such action by the United Nations and other U.N. specialized agencies:

> It can be said that Israel is in many respects alone in the world because its own orientation is against the general trend. Even American-Israeli relations are in the process of change, owing largely to the fact that the United States can no longer centre its Middle Eastern policies on the interests of Zionism alone.[50]

If the physical and mental isolation of Israel from its neighbors and from much of the outside world continues and the Arabs continue to gain their power, prestige, and credibility there is a corresponding loss of Israeli position in the world.

What was the energy crisis for the industrialized world was for the oil producers tremendous political and economic gains. The relative Arab military success in the October war and their control of the use of oil as a weapon showed that they were acquiring not only the tactics of modern warfare but also the art of diplomacy. Moderation has been constantly facing pressure from those in the Mideast who feel from experience that militancy pays.[51] It was for this reason that Saudi Arabia shunned its policy of moderation and decided to use the oil weapon in the Mideast conflict. It satisfied the nationalist aspirations of the Arabs. At the same time, Saudi Arabia and other Arab nations realized that their weapon was a two-edged sword. "Oil is a doubled-edged weapon,"[52] said Habib Bourguiba, president of Tunisia. Before the weapon could cause irreparable damage to the American economy which, in turn, would affect Arab economic interests, the oil embargo was eased. As one Arab minister contended:

> If the embargo were to remain, we would see a major recession in America. That, in turn, would affect all of us adversely. Our

economies, our regimes, our very survival depend on a healthy U.S. economy.[53]

The early lifting of the embargo indicated that the Arabs, particularly the Saudi Arabs, decided not to abuse their economic power. "There is a growing realization that this power—based on hydrocarbons—is finite and that alternative sources of revenue should be developed speedily."[54] According to present indications Saudi Arabia's oil reserves will dry up in the year 2039, at which time the desert kingdom itself will be an oilless country.[55]

Saudi Arabia's oil resources were developed exclusively by the Americans. It is obvious that the Saudis need American resources of technology to set up a functional economic infrastructure on which a future oilless economy could be based. This is a gigantic task and, judging from the available pool of resources of technology in the world, can only be delivered by the Americans, Europeans, and Japanese in a massive way to the Saudis.

Again, speaking from the economic point of view, the oil embargo has paradoxically strengthened the American position in the Middle East, especially in Saudi Arabia. Saudi relations with the American oil companies have undergone many changes since the beginning of their operations in the 1930s, but it is noteworthy that, although the principal object of the oil embargo was the United States, American oil companies still continue to operate in Saudi Arabia.

The coming of the energy crisis and the increasing dependence of the world on Middle Eastern oil have made oil a controversial subject and have created a climate full of doubt and suspicion regarding the motives of the host government and its relationship with the oil companies and the companies' relations with their own countries.

While many of the host governments in the Arab world have either nationalized or signed participation agreements with the oil companies, particularly during the periods preceding and following the oil embargo of 1973, restraint and moderation have marked the behavior of Saudi decision makers. In the oil industry there is a feeling that American oil companies in Saudi Arabia will wind up with arrangements similar to those received from Iran in early 1973. Iran nationalized the companies and granted them lucrative contracts to continue to operate as before and to buy oil at the prevailing prices. Or, perhaps, Saudi Arabia may establish control from within, that is, participation with the oil firms by the formation of joint ventures.

On the other hand, the oil firms, in defense of the high earnings they make on their investments in Saudi Arabia and the Middle East,[56] argue that their extra profits are eventually eliminated in downstream activities, such as transportation, refining, and marketing, on which they make very little or no profits at all. Hence, they say, the profitability of their total investments abroad is not excessively higher than other investments in foreign countries.

Saudi Arabia and other oil-producing states do not accept this argument. To them, the problem of the low earnings on downstream investments does not change the fact that income per dollar invested upstream in the Middle East is usually high in comparison with other industries and other regions. Furthermore, they believe that the high profits earned upstream are transferred abroad to finance less profitable investments in other areas, coupled with devaluation and inflation of the dollar and other hard currencies in the West, and the search for new sources of crude and other energy in the West, which would eventually compete with the existing Middle Eastern sources from which these profits accrue. In other words, the profits made out of the sale of Arab oil are used to finance the diversification program of the oil firms, which might reduce the importance of Arab oil in the future, and to compensate the firms for low profits made in the downstream operations which are conducted mainly in the industrialized and more developed states of Europe and North America.

Economically, Saudi Arabia wants to maintain a cooperative relationship with the oil companies and the industrial world since her freedom to act individually is severely restricted by the inability to carry out the responsibility for oil operations. Politically speaking, the changing relationship between Saudi Arabia and the United States is not a function of the political situation arising from the Arab-Israeli conflict, rather it is a function of the Saudi perception of the energy crisis and the oil kingdom's special role in solving it.

The scale of the problem is of course too great to be solved by an expression of goodwill and too complex to be solved merely by bilateral agreements with individual oil-consuming nations. Meanwhile, both in dialogue and in action, an appreciation in Saudi Arabia of the power and responsibilities of wealth are apparent. A sense of collaboration to overcome the problem arising from the world energy crisis and a shift in world debtor-creditor relationships caused by the accumulation of wealth appear to be present in the minds of Saudi leaders.

A great deal of the discussion revolves around the scale of the revenue surpluses being amassed by Saudi Arabia. By 1980, if peace prevails in the world, Saudi Arabia will probably be producing 20 million barrels of oil a day and could possess $200 billion.[57] This enormous accumulation of funds will not only bring power but will enhance Arab prestige in the world. On the other hand, Saudi Arabia and other deserts of affluence in the vast land of poverty will not only generate tension between the oil haves and have-nots in the region, but will create rivalry between the two oil giants—Saudi Arabia and Iran—which could ultimately induce subversive elements, both internally and externally, to disrupt the status quo.

Even so, its new stature is such that few states in the region can afford to antagonize Saudi Arabia. Confidently it counters Iraq, Syria, Libya, and the Popular Democratic Republic of Yemen. Cooperation with Iran has been reasonably good so far, if limited. This relationship, however, may be undergoing strain from the dispute over oil pricing and it could deteriorate if Iran goes too far in making the gulf an Iranian lake and further expands its military presence in Oman.

On the personal level, there is little or no problem between the Saudi and Iranian rulers on the one hand, and the Western governments on the other hand. But in the ordinary Arab's eyes, the West is interested only in the Middle East's oil, wanting to dominate politically and militarily so that it can exploit the region economically. In the past, "To this end, Western powers bribed or bullied the Arab ruling classes into collaboration and the people remained poor and oppressed."[58]

The oil embargo was established not to effect a cease-fire in the October war but to seek explicitly the evacuation of Israeli forces from all Arab territories occupied during the June war of 1967. This indicated that the Arabs were willing to accommodate Israel within the original boundary given to it by the United Nations in 1948. Another political reason for which the oil embargo was imposed was to restore the "legitimate rights" of the Palestinian people. Although it was not spelled out during the October war, the restoration of the legitimate rights of the Palestinians meant their return to an independent Palestine which, in turn, implied its liberation from Israeli occupation. As subsequent events showed, this was what the Arabs actually meant by the restoration of the legitimate rights of the Palestinians.

The Arab-Israeli conflict is essentially a Palestinian-Israeli issue, due to the fact that the Palestinians who were driven out of their homeland in 1948 live as second-class refugees scattered all over the Arab world. Although they are Arabs, many Palestinians prefer to be known as Palestinians due to their nostalgic feeling toward their homeland.

The Palestinian movement was dramatized at the Rabat summit meeting of the Arabs in October 1974, where the PLO won a major diplomatic victory when 20 Arab nations backed its claim as the sole representative of the Palestinian people, despite King Hussein's opposition. He, however, succumbed to the unanimous pressure of the Arab world.

The United Nations, which in effect drove the Palestinians out of their homeland by a 1947 resolution that created Israel, bent its traditions by inviting the head of PLO, a political movement, to address the General Assembly as though he represented a government.[59]

Yasser Arafat, the head of PLO, while addressing the United Nations on November 13, 1974, took an all-or-nothing approach which culminated in the passage of a resolution on November 22 by the General Assembly recognizing the rights of the Palestinians to independence and sovereignty in Palestine.[60] The General Assembly voted the resolution by 89 to 8 with 37 abstentions, mostly Europeans and Latin Americans. The United States, Israel, Norway, Iceland, Bolivia, Chile, Costa Rica, and Nicaragua voted against the resolution.[61]

Of all the developments in the Middle East since the creation of Israel, the passage of the Palestine resolution is most significant. It is so because the United Nations which in a sense created the State of

Israel by a similar resolution, has passed this counterresolution to undo Israel and to recreate Palestine. In a simplistic sense Palestine has achieved its international personality, being endorsed by the world body in a landslide vote. What remains to be done to restore this duly established legitimate right in Palestine is, however, most difficult to accomplish. As happened in the case of Israel, a U.N. resolution to partition Palestine among Jewish, Arab, and international sectors was implemented by the use of brutal force (the Jewish sector was provided with 56 percent of the territory, the Arabs were given 42 percent, and the remaining 2 percent in Jerusalem was internationalized) and, in the process, Israel took over the whole of Palestine by successive wars. It would seem that the Palestinians can regain their lost territory only by applying the same technique.

The aftermath of the use of oil as a weapon of diplomacy has accomplished the peaceful aspect of the recreation of the State of Palestine. The case of Palestine was strengthened by an unexpected announcement when the Shah of Iran, who has shown new support for the Arabs, joined President Sadat in calling for Israel's total withdrawal from occupied Arab land by affirming the right of the Palestinian people in their homeland.[62] If the current trend in Iranian policy to befriend its Arab neighbors including Iraq continues, Israel's secure supply of oil will be threatened in case of another Arab-Israeli war, which would be most dangerous for Israel's survival as it has no other source of oil supply in the region. In pursuance of his policy of reestablishing the Persian Empire, the Shah of Iran may join the next Arab war efforts against Israel which may ultimately enable him to play the role of policeman in the gulf.

As time passed, the effect of Arab petrodiplomacy was felt more in the international economic field than in the political arena, which happened as a result of the oil price escalation for which Saudi Arabia can be blamed very little. Although Saudi Arabia led the Arab nations in imposing the oil embargo, probably there was no intention of increasing the price of oil suddenly. Nevertheless, soon the situation went out of control and, taking advantage of the oil embargo, Iran and Venezuela took the lead in escalating the price. The quadrupling of oil prices in little more than a year is perceived to be the cause of economic chaos in the world today, although this is not the whole cause of all economic ills in the world. Yet, the sudden fourfold increase in oil prices, almost at will, by the OPEC nations and the back-door method by which it was done has caused the greatest disarray in the world today.

The growing disparity between oil haves and have-nots has been conducive to international financial chaos and misunderstanding. The inequality has been accentuated by the fact that international oil is now owned by a few developing nations which are still poor compared to the standard in the First World. Naturally, they jealously guard their newfound wealth against encroachment from the outside. The simple fact is that not all jealousies and conflicts are amenable to quick solution.

Even though OPEC does not see it this way, no return to the old order is possible. The transfer of financial power with all its components has already taken place and the world has to adjust to the new reality, as the Arabs call it. The only danger is that of extremism and clearly the imposition of an oil embargo in 1973 by OAPEC and the subsequent explosion in prices of oil by the OPEC nations constituted extreme actions.

Recognizing the dimension of the international financial crisis and the weakness in their oil policy OPEC nations offered $9.6 billion to the special fund created by the IMF for recycling nearly all of this amount to the Fourth World countries primarily for payment of their petrobills.[63] The Arab oil nations, in addition to their commitment in association with OPEC, are compensating poorer Arab countries for higher prices of oil. They have offered $80 million as compensation to the Arab oil have-nots.[64]

The oil producers blame inflation and past colonial exploitation as the most important contributing factors in the oil price hike; they also continue to argue in favor of escalating the price further. Saudi Arabia differs with Iran, Algeria, Libya, and Iraq on the question of further price increases. King Faisal decided not to attend the first meeting of the heads of state of oil-exporting countries in March 1975, as they had reportedly refused to effect a reduction in oil prices. The Saudis also refused to consider proposals by Iran, Algeria, and other oil exporters that there be a production quota for each of the 13 OPEC nations as a means of sharing production cutbacks resulting from a drop in demand from industrial countries.

As long as OPEC unity is maintained it can do whatever it likes. Historically speaking, all international cartels contained the seeds of their own destruction. OPEC cannot be an exception to this. Already the divergencies in the interests of its members are evident. As producers are many, some competitive forces may operate to break the unity. Saudi Arabia, having achieved recognition of its position as an oil superpower, seems to have been developing greater confidence and responsibility in its international behavior, except in its attitude toward international Zionists whom the oil kingdom discriminates against as far as participation in Saudi development is concerned. Since the lifting of the oil embargo Saudi Arabia feels no economic pinch and may want to adjust the price of its oil downward. Indeed, Saudi Arabia has been advised by the Stanford Research Institute, hired by the kingdom, that its long-term interests would be better served by a lower price of oil.[65] King Faisal made a public commitment to lower oil prices, and he was believed to be working on that, as his boycott of the first summit of the oil-exporting nations seemed to indicate. King Khalid has announced that he would follow the policy of his brother, the late King Faisal. It is under Saudi Arabian influence as the largest producer of oil within OPEC that the price of oil for one and a half years remained steady since the lifting of the embargo until it was raised by 10 percent in the fall of 1975.

The UAE has blamed other OPEC nations for lack of coordination in oil production schedules. The decrease in production has resulted in the loss of earnings in the UAE. Iran, Kuwait, and Algeria are urging other OPEC members to cut production to keep prices up. This is seen as a counter move to the drops in consumption by the industrialized world which have, in turn, created an oil surplus, placing a natural downward pressure on prices.

Because of the factiousness among the Arab states, they never had any unified military action against Israel. Even during the 1973 war only two confrontation states fought directly against Israel. Petrodiplomacy was the only weapon the Arabs used more unitedly. The oil embargo had a limited success as far as the Israeli withdrawal from the west side of the Suez Canal and part of the Golan Heights was concerned. It is to this aspect of the Arab-Israeli conflict and the U.S. efforts at settlement that we turn in the next section.

U.S. Efforts At Peace Mediation

Initially, the U.S. policy in the Middle East centered on containment of the Soviet Union. Directly related to this policy was the maintenance of stability and peace in the Mideast as lack of them was considered conducive to the growth of Soviet influence. Another major U.S. policy interest was the support of Israel. As detente was reached between the two superpowers, the U.S. policy in the Middle East began to change. "Containment and the fear of a monolithic communism have outlived their utility. The Dulles-devised regional security pacts of the fifties are remnants of a polarization that is giving way to pluralism."[66] Politically speaking, the current U.S. policy in the Mideast is based on gaining the friendship of as many nations as possible and at the same time ensuring Israel's security. Of the two major principles of U.S. policy in the Middle East, the second interest, that is, support for Israel, has become more important. This is because Israel, Russia, the Zionists, the Arabs, the Third and Fourth Worlders—all for their different reasons—have been able to identify the United States with Israel.

While Arab-Israeli conflicts have their own causes and characters, they are interrelated when they involve the interests of the superpowers. Apparently, the Soviet policy vis-a-vis Israel is not aimed at the destruction of the state but, through pressure, to make it give up its gains. This was the philosophy of Security Council Resolution 242 of November 22, 1967, that Washington and Moscow helped to draft. At one point during the fourth Arab-Israeli war the two superpowers were about to engage in armed conflict in the Middle East but, thanks to American strategy and the famous shuttle diplomacy of the U.S. secretary of state, the dangerous confrontation was avoided in the spirit of detente. In the words of Nadav Safran:

> Besides defusing an explosive situation, starting off a process of negotiation for the first time in 25 years, setting up hopeful precedents for compromise, and beginning to generate mutual trust between the antagonists, the disengagement agreements involved a substantial modification of past patterns of inter-Arab politics.[67]

The other key Security Council resolution on the subject is number 338 of October 22, 1973, in which Kissinger diligently incorporated provisions making Resolution 242 mandatory. Taken together these resolutions require the parties to negotiate peace agreements "concurrently" with security arrangements. According to an interpretation by Eugene V. Rostow, a former U.S. under secretary of state, "Under the Security Council Resolutions, Israel is not required to withdraw one inch from the territories it holds as the occupying power until its Arab neighbors have made peace."[68]

Following the end of war in 1973 both the radical and the moderate leaders of the Arab world began to cooperate with the United States in reaching the disengagement agreements. This rapprochement between the United States and the Arab nations placed the United States in a position of trust and influence in much of the Arab world.[69] The elements that made this change in the Arab attitude possible go further than the disengagement agreements. Again in the words of Safran:

> They include strong shared interests with the United States on the part of conservative Arab countries, beneath their expressed resentment of American policy toward Israel; strong resentment on the part of radical Arab countries toward the Soviet Union, beneath their cooperation with it; fears on the part of both radical and conservative Arab countries about the costs and outcome of continuing confrontation and war; and hopes on the part of the Arabs that American behavior in the last stage of the October War, which saved them from the humiliation of another defeat, portended a favorable change in American policy.[70]

Kissinger's unilateral personalized shuttle efforts to nudge the Mideast disputants toward a step-by-step settlement received a great setback following the Arab summit meeting at Rabat, which strengthened the role of PLO, and the U.N. resolution which created a homeland for the Palestinians, embracing the west bank of the Jordan River and Israel itself. The situation is further complicated as the oil-rich nations of the region are engaged in huge arms-purchasing deals and in supporting the PLO with necessary funds. According th George W. Ball, a former U.S. under-secretary of state:

> One lesson we should have learned from the experience of past months is that highly personalized diplomacy is effective

> only in a bilateral setting; it has limited value in a complex situation involving many countries. Thus the attempt to settle the Arab-Israeli issue by shutting out both the more activist Arab states and the Soviet Union was predestined to failure.[71]

From the beginning of his shuttle diplomacy it has been clear that the Kissinger approach has been pragmatic and not legalistic; however, it has not been clear as to how he can secure an Israeli withdrawal within the boundary which existed before the 1967 Arab-Israeli war. Analyst Ball, among others, believes that this could be accomplished "only in the multilateral setting of the Geneva Conference, with participation of all the principal Arab states, including the most radical, and with the Russians acting as co-chairman."[72]

The implicit assumption is that Israel would be required to return to its 1967 frontiers, with only slight modifications. While this seems unacceptable to Israel, "Kissinger is understood to feel that if the United States were ready to press for a full Israeli pullback there would be no need for Soviet cooperation in the process."[73] Kissinger's rush in his shuttle diplomacy was done purposely in order to keep the momentum of negotiation going. However, the momentum stopped for some time and this strategy came to a halt largely because of the changing role of PLO in the settlement of the Arab-Israeli conflict. The legitimacy of PLO cannot be denied now. Kissinger's failure to promote some sort of dialogue between Jordan and Israel in advance of the Rabat summit was a great opportunity missed. Previously the Arab nations as a whole demanded the destruction of Israel. While this stand was modified, they still wanted Israel to withdraw from the territories it occupied in 1967 and 1973. PLO leader Arafat reaffirmed before the U.N. General Assembly in November 1974 that the creation of a secular Palestinian state in Palestine, which includes Israel proper, was the goal for which the Palestinians were striving. So far the Arabs have fought the Palestinians' wars. Now the Palestinians themselves are emerging in a leading role in an attempt to liberate their homeland from Jewish occupation. It was precisely for this reason that PLO was given sole responsibility at Rabat to speak and act for the Palestinians.

Just as the United States finds it politically impossible to wash its military hands of Israel, so the Arabs find it politically suicidal for their policy not to fight for the Palestinians against Israel. Otherwise, both Egypt and Syria would probably be more than happy to settle for the return of their territories occupied by Israel. If this were the case, now that Russian interest does not lie in opposing U.S. policy, a settlement of the Mideast impasse would be much easier to reach.

From the beginning of his deinstitutionalized, step-by-step diplomatic approach Kissinger has been warned by many analysts of the crippling contradiction inherent in his policy. Whether right or wrong, Kissinger's followup moves and success or failure of his diplomacy will be seen in time. He has tried repeatedly to convince President Sadat of the merits of his approach to peacemaking in the Middle East, while the

Syrians and the Palestinians denounced the move. The Sadat-Kissinger understanding was based on their conviction that once Israeli forces pulled back from the strategic Mitla and Giddi mountain passes in the Sinai Peninsula as well as the Abu Rudeis oil fields at the southern tip of the Suez Canal, peace mediation efforts in other fronts could be undertaken at the Geneva Conference. Israel agreed to the return of these territories and the oil fields in exchange for a declaration of nonbelligerency from Egypt, which President Sadat was not willing to make directly, but for an explicit assurance of supply of oil in case of emergency and the stationing of 200 American civilian technicians to man early-warning stations in the Sinai Peninsula. Sadat agreed to make an implicit pledge not to resort to the use of force. It was reported that the Egyptians were ready to meet many of the components of military and economic sides of nonbelligerency. But Israel wanted a categorical declaration of political nonbelligerency. The Israeli intransigence is based on the ancient and still practiced concept that war is a legal means of acquiring territory, and it should not be returned to its former owner without a declaration of political nonbelligerency. Egypt could probably agree to such a formal declaration if all the Arab territories occupied by Israel were returned at the same time, or in other words, if the end result preceded the beginning endeavor. The Egyptian attitude was that a declaration of nonbelligerency would be almost tantamount to a declaration of illegality of war by Egypt, while Israel keeps most of the occupied territories in her possession. By agreeing to a declaration of nonbelligerency, Egypt would consent to the legitimization of Israeli occupation of the rest of the Arab territories.

Egypt also knew that once it agreed to an interim settlement with Israel this would be calculated to arouse deep division in Arab ranks. The centrifugal forces would begin to reassert in Arab politics, and Palestinian activists might resort to violence on a larger scale. As soon as the news of the beginning of the last (eleventh) round of Kissinger's step-by-step approach broke, the Palestinians moved into open conflict with President Sadat. President Assad of Syria, being fearful that a separate Egyptian settlement with Israel might diminish the prospect of getting back the Golan Heights through peaceful means, created a united political and military command with the Palestinians.

Needless to say a peaceful settlement in the Mideast depends on complete Israeli withdrawal from the whole of the Sinai Peninsula and the Gaza strip, including the former Egyptian outpost of Sharm el Sheikh, which commands the Israeli entrance to the Gulf of Aqaba as well as the Gulf of Suez. Sharm el Sheikh is important for Israel's entry into its only Red Sea port, Elat. Israel also insists that it must retain the Golan Heights and the Nahal outposts along the Jordan River for security reasons. Above all, Israeli annexation of Jerusalem, where the former is establishing many permanent settlements, poses another seemingly insoluble problem. That which is a security hazard for Israel is equally a security problem for the Arabs and, after all, these are Arab and Palestinian territories occupied by Israel.

Under these conditions, what is the prospect of a peaceful settlement? The hope now is to reconvene the Geneva Conference on the basis of Resolution 242 under the auspices of the United Nations, the United States, and the Soviet Union in which Arab nations, Israel, and the Palestinians are supposed to participate. The planning and organization of a conference of so many diverse elements, involving issues as complex as Israel's borders with Syria, Jordan, and Egypt, and the Palestinian problem is such that no quick session nor quick decision can be expected. Even assuming that the conference is resumed in Geneva, endless debates would be sure to build up pressures on Israel to withdraw its forces from all occupied Arab territories. The contemporary American temper is apparently evenhandedness in the U.S. Mideast policy. Not only are some U.S. senators and congressmen becoming more discriminating and reserved in their attitudes toward Israel but even President Ford and Secretary Kissinger have complained of Israeli intransigence and inflexibility.[74] On top of this President Ford has ordered a formal review of U.S. policy in the Mideast. The White House announcement came at the height of the collapse of the tenth round of American shuttle diplomacy when the world was blaming Israel for its uncompromising attitude. Probably this was done to prevent the cooling of otherwise warm U.S. relations with such moderate Arab states as Egypt and Saudi Arabia. But nothing spectacular can be expected of the reassessment of American policy in the Mideast. However, the Jewish influence in U.S. policy in the Middle East is meeting some opposition and at the same time an opinion favoring contact with Arabs to appreciate their point of view is appearing. Israel's problem is with the White House, not so much with the Congress. In this, Israeli strategy is to stand firm, to keep up the lobbying activities, and wait out American administration's reassessment of its policy toward the Middle East on the assumption that the White House cannot ignore Israeli request of $2.6 billion in military and economic assistance unless Congress is prepared to back President Ford. Therefore, Israel, which has agreed to a limited pullback on the Egyptian front, after fifteen commitments made to Israel by Secretary Kissinger as part of the Sinai agreement, is not likely to yield to pressures for withdrawal from all occupied lands. This is because Israel insists that there should be a step-for-step approach for peace settlement or agreement-to-make-agreement as the Security Council Resolution 242 and 338 seem to prescribe. Under this circumstance, the Geneva Conference, if convened, could very well collapse in chaotic disagreement, or prolonged discussion could follow, virtually guaranteeing a potentially explosive deadlock. To make the point differently, the success of the Geneva Conference will be judged by the postponement of war rather than by the progress of discussions toward peace. It is believed that a no-war no-peace situation, therefore, will prevail in the Arab-Israeli zone for some time. However, the Sinai agreement has created an illusion of peace in the Mideast. The Russians have suffered another setback in the region and President Sadat has turned increasingly toward the United States for economic and military assistance.

Since the 1973 Arab-Israeli war, Secretary of State Kissinger has made seven trips to the Mideast in search of a framework for an agreement between Egypt and Israel. However, the irreconcilable differences between them could not be reconciled. The fragility of earlier American peace efforts in Southeast Asia demonstrated how transient success on this kind of a mission can be. This came at a time when reversals and setbacks to U.S. policy from Southeast Asia to Southern Europe to the Mideast dominated the world news. Further, Kissinger's approach was to separate and distinguish the Arab-Israeli conflict from the Arab-Palestinian-Israeli dispute by isolating and ignoring the Palestinians. With the passage of time the twin Arab-Palestinian issue has become inseparable and indistinguishable; it can no longer be compartmentalized. The issue of Palestine is the original problem in the Middle East; the Arab-Israeli conflict is a derivative of it. But Kissinger's policy seemed to be based on the theory that an uprooted tree can be brought back to life by watering its leaves.

If the above analysis is correct, the Palestinians who have not succumbed to the pressure in the past, will be reinforced and emboldened by the lack of resource constraints in the oil-rich Arab states. The oil-short Arab states, such as Egypt and Syria, may augment the Palestinian's aspirations to return to their homeland by contributing their military might. A large-scale Palestine-Arab-Israeli war opening at the Syrian, Egyptian, and Lebanese fronts could develop in the Middle East—not immediately but in ensuing years. It is difficult to predict the outcome of such a war now. However, if American support of Israel continues, as seems likely because of the policy reiteration by as many as 76 U.S. senators on May 22, 1975,[75] the Arabs will in all probability lose the war again and the Palestinians may still have to wait a long time to regain their lost homeland.

NOTES

1. Shaykh Ahmad Zaki Yamani, "Prospects for Cooperation Between Oil Producers, Marketers and Consumers: The Issue of Participation and After," World Energy Demands and the Middle East (Papers delivered at the 26th Annual Conference of the Middle East Institute, Washington, D.C., September 29-30, 1972), p. 96
2. Tareq Y. Ismael, The Middle East in World Politics: A Study in Contemporary International Relations (Syracuse: Syracuse University Press, 1974), p. 236.
3. Middle East Economic Survey, June 15, 1974.
4. George Lenczowski, Oil and State in the Middle East (Ithaca, N.Y.: Cornell University Press, 1960), p. 188.
5. Ibid.
6. Ibid.

7. Leonard Mosley, *Power Play: Oil in the Middle East* (Baltimore: Penguin Books, 1974), p. 343.

8. Ibid.

9. Ibid., p. 344.

10. *Middle East Economic Survey*, August 11, 1972.

11. Quoted in *Middle East Economic Survey*, July 21, 1967.

12. Ibid.

13. *U.S. News and World Report*, January 15, 1973.

14. Malcolm H. Kerr, *The Arab Cold War* (London: Oxford University Press, 1971), p. 129.

15. Nadav Safran, "The War and the Future of the Arab-Israeli Conflict," *Foreign Affairs*, January 1974, p. 219.

16. See an English review of the original study in Arabic in *Journal of Palestine Studies*, Autumn 1973, pp. 142-44.

17. *Washington Post*, June 17, 1973.

18. *Christian Science Monitor*, September 4, 1973.

19. Mosley, op. cit., p. 423.

20. Safran, op, cit., p. 220.

21. Ibid., p. 221.

22. Dana Adams Schmidt, *Armageddon in the Middle East* (New York: John Day Co., 1974), p. 212.

23. Ibid.

24. *Economist* (London), November 3, 1973.

25. *Daily Telegraph* (London), November 17, 1973.

26. *Sunday Times* (London), November 14, 1973.

27. Schmidt, op. cit., p. 213.

28. John H. Lichtblau, "Arab Oil and a Settlement of the Middle East Conflict," in *After the Settlement: New Direction and New Relationships* (Papers delivered at the 28th Annual Conference of the Middle East Institute, Washington, D.C., October 12, 1974), p. 1.

29. Ibid., p. 5.

30. Ibid., p. 6.

31. *Washington Post*, December 24, 1973.

32. Lichtblau, loc. cit.

33. Ibid., p. 5.

34. Adapted from Joseph Kraft, "The Rising Price of Oil," *Washington Post*, December 1, 1974.

35. *Washington Post*, December 31, 1973, and New York *Times*, December 31, 1973.

36. Ibid.

37. Ibid.

38. *Washington Post*, December 31, 1973.

39. *Washington Post*, December 26, 1973.

40. New York *Times*, October 18, 1973.

41. *Middle East Economic Digest*, March 22, 1974.

42. *Washington Post*, March 28, 1974.

43. Lichtblau, op. cit., p. 5.

44. *Newsweek*, March 25, 1974.

45. Ibid.

46. *Arab Oil and Gas Journal*, May 1, 1974.

47. Ibid.

48. Schmidt, op. cit., p. 216.

49. *Washington Post*, April 2, 1975.

50. Alan R. Taylor, "The Isolation of Israel," *Journal of Palestine Studies*, Autumn 1974, p. 93.

51. *Middle East Economic Digest*, March 22, 1974.

52. Quoted in Mosley, op. cit., p. 426.

53. Quoted in *Newsweek*, March 25, 1974.

54. *Middle East Economic Digest*, December 28, 1973.

55. Mahdi al-Bazzaz, "Middle East Oil Revenues: An Assessment of Their Size and Uses," *Middle East Economic Digest*, March 15, 1974.

56. *Newsweek*, February 4, 1974. It was reported that the oil companies' profits for 1973 over 1972 were as follows: Exxon, 59%; Mobil, 47%; Texaco, 45%; and Standard of California, 40%. A good share of the 1973 profits were, as maintained by the companies, the result of the big boost from the devaluation of the U.S. dollar, which allowed the companies to write up the value of their other currencies by huge amounts.

57. Edward R. F. Sheehan, "Mastermind of Mideast Oil," *Reader's Digest*, September 1974, p. 84.

58. *Newsweek*, October 28, 1974.

59. *New York Times*, November 16, 1974.

60. *Washington Post*, November 23, 1974.

61. *United Nations General Assembly Provisional Verbatim Record No. A/PV. 2296* (New York, November 22, 1974), p. 46.

62. *Washington Post*, January 12, 1975.

63. *Washingtonian*, March 1975.

64. *New York Times*, May 5, 1975.

65. *Fortune*, May 1975, p. 280.

66. Abdul A. Said, ed., *America's World Role in the 70s* (Englewood Cliffs, N.J.: Prentice-Hall, 1970), p. 11.

67. Nadav Safran, "Engagement in the Middle East," *Foreign Affairs*, October 1974, pp. 49-50.

68. Eugene V. Rostow, "A Basis for Peace," *The New Republic*, April 5, 1975, pp. 12-13.

69. Safran, op. cit., p. 54.

70. Ibid., p. 56.

71. George W. Ball, "The Looming War in the Middle East and How to Avert It," *Atlantic*, January 1975, pp. 10-11.

72. Ibid., p. 6.

73. *Washington Post*, December 22, 1974.

74. *Washington Post*, March 28, 1975.

75. See *Washington Post*, May 23, 1974.

CHAPTER

7

CONTINUING SIGNIFICANCE OF THE EMBARGO OPTION

Although it would be a much more serious decision this time than it was in 1973, most observers maintain that it is almost axiomatic that a renewed Arab-Israeli war would immediately precipitate a more severe oil embargo. It is also axiomatic that the United States would not permit Israel to be wiped out in a war. By the same token Russian support of the Arab cause is not axiomatic. Russia backed out of its support of the Arab war efforts in 1967, and again in 1973 it did not fulfill its pronounced policy of sending volunteers to the Middle East to protect Arab interests as soon as the United States put its strategic forces on alert all over the world. In fact, modern Russian history shows that the Soviet Union does not fight other people's wars except by helping them with arms and training and by extending diplomatic support at international forums. At the same time, whenever possible, the Russians have taken over other lands, as it did in Iran, to give a Middle Eastern example only. In contrast, direct U.S. involvement in other people's wars as occurred in Vietnam as well as the American armed intervention in a local conflict as happened in Lebanon, is still fresh in the world's memory.

As a result of detente the two superpowers may come to terms for scoring gains in the new economic game now in progress in the Middle East. The name of this new economic game is oil. "Oil is the most international of all problems of the Middle East. . ."[1] The two superpowers, the industrialized nations of both Eastern and Western Europe, Japan, and the preindustrialized and the nonindustrialized countries are all interested in securing oil supplies at affordable prices. The oil producers are aware that their drive for higher prices for oil does not evoke much sympathy among those nations which pay the bills. And these include all nations except the surplus oil producers. By imposing an embargo on a vital commodity like oil, OAPEC turned the Middle East crisis into a global trauma. Even the Communist nations and a few Fourth World countries, the main supporters of the oil price increase, are feeling the pinch of the energy crisis.

"The Soviet Union has recently taken an unexpected interest in Middle Eastern oil."[2] Historically speaking, even in 1921 the Soviet Union attempted to establish its claim on northern Iranian oil. During the interwar years Russia joined the Western powers in obtaining oil concessions in the Middle East. In 1947, Stalin extracted the promise of oil concession in Iran as his condition for withdrawing Russian troops from Azerbaijan.[3] But the Russian attempts failed mainly because of U.S. and British interference. During the decades of the 1950s and 1960s the Soviet Union became an oil surplus country. Although reliable statistics from the Soviet Union are not forthcoming, knowledgeable experts believe that consumption of oil is exceeding domestic production at an extremely rapid rate.[4] New oil discoveries have been made in Siberia, but the "demand for oil will, in 1980, exceed domestic supply by about 100 million tons a year. . ."[5] By the same year Eastern Europe may require about 170 million tons of oil per year, but total production will be about 30 million tons.[6]

American and Western interests in Middle Eastern oil have been thoroughly described previously so that further discussion here is not deemed necessary to put those interests in perspective. Suffice it to say that the interests of the two superpowers in the oil wealth of the Middle East are in conformity with each other—both they and their friends need badly the liquid gold of the Mideast for continued progress during the decade beginning in 1975. In about 1985, as discussed previously, the importance of oil for continued survival of modern civilization will begin to be diminished. The decade 1975-85 is, nevertheless, absolutely crucial.

Diplomacy has had a chance in the settlement of the Mideast crisis, and obviously its significance is not completely lost. Also it can still succeed, provided the Palestinians give up their policy of calling for Israel's destruction and agree to settle for a homeland on the west bank of the Jordan River and the Gaza strip. A growing number of specialists maintain that unless peace is established soon, war could come in the Middle East, possibly during the next few years. Shuttle diplomacy is already not as effective as it was soon after the war ended in 1973. In the words of Kissinger himself, "There was a time for shuttle diplomacy, and there is a time for quiet diplomacy."[7] Quiet diplomacy is, seemingly, a process of negotiation from a position of helplessness. The U.S. helplessness emanates from its failure to stop the oil price increase, to guarantee the flow of Mideast oil to Western nations, and to stop the "tyranny of the majority" at the United Nations from admitting PLO as an observer and from depriving South Africa of its seat in the twenty ninth General Assembly Session.[8] Coincidentally, the same majority in the U.N. Educational, Scientific, and Cultural Organization cut off support for Israel on the ground that it is altering the physical and cultural character of Jerusalem. The situation is described as follows:

> . . . with the Palestinians playing a newly pivotal role in intra-Arab politics and with the Arab world as a whole seemingly convinced that oil will turn the tables on Israel, Rabat and its aftermath seem to have made new war much more likely. It could break out in a number of ways, ranging from Arab attacks in response to Israeli rejections of demands for new withdrawals from the occupied territories to an Israeli preemptive war triggered by circumstances such as a new Arab "war of attrition," evidence of a coming Arab attack, or simply a feeling that a blow must be struck before the tables are hopelessly turned.[9]

In this situation Israel would probably appear as the attacker, and Israel cannot go to war unblessed. The blessing will have to come from the United States. The question is, will the United States give its blessing to Israeli war efforts? The United States is faced with two options: either to help Israel fight its war or to fight the Israeli war itself. In the past the United States has helped Israel to fight its wars with the Arabs through arms aid and diplomatic support. In the future the United States is also likely to support the Israeli war cause against the Arabs.

Assuming that the United States is able to persuade Israel not to wage a preemptive attack against the Arabs, the next question is whether the Arabs and the Palestinians will keep quiet while their territories are under Israeli occupation. The answer is no, as:

> Processes of modernization throughout the Arab world have considerably increased Arab abilities to fight a sustained war. And the decisive change in the economic balance between Israel and the Arab states brought about by the four-fold increase in oil prices . . . has removed all resource constraints from Arab armament efforts.[10]

AMERICAN INVASION OF ARAB OIL FIELDS

The possibility of another major war cannot be ruled out in the Mideast, and it could precipitate an energy "Pearl Harbor" for the West. In that event "the United States might expect pressure for an American military intervention from many who are already urging that we seize the oil production of the Middle East to save the West from bankruptcy."[11] The possibility of American intervention with military force to take over Arab oil production facilities in order to assure a steady supply of reasonably priced oil to the Western markets is perhaps real.

The threat of American intervention in the Middle East is believed to be real because U.S. military troops have undertaken desert maneuvers for the first time in 40 years. "American Army and Marine units in the U.S. have stepped up desert-warfare training. . . . For its part, the Navy sailed the aircraft carrier Constellation into the Persian Gulf."[12] Although there is disagreement among State Department and Pentagon policy planners, the latter is keeping its contingency plans up-to-date and the "petroland" maneuvers, which began in 1973, seem to be continuing in full swing. Mideast observers make a connection between this and U.S. energy needs.[13]

The justification for American invasion of the Mideast, aside from protection of Israel, is that depriving a nation of access to basic resources has been historically considered a causus belli. In his speech to the World Energy Conference in Detroit on September 23, 1974, President Ford echoed: "Throughout history, nations have gone to war over natural advantages such as water, food or convenient passages on land or seas."[14] To be meaningful, it is assumed that such intervention will be directed in a massive way against Saudi Arabia, the largest oil exporter.

The United States has made a contingency plan ready for an expedition in the Arab world. This invasion plan, drawn up by the Pentagon, was revealed by a Washington-based professor and defense consultant who wrote an article under his pseudonym entitled "Seizing Arab Oil: A Blueprint for Fast and Effective Action," which was published in the March 1975 issue of Harper's.[15] In his article, he discussed, inter alia, the U.S. invasion plan, code-named Dhahran Option Four, which had been approved by the U.S. National Security Council (headed by the U.S. president himself) before Kissinger, in an interview with Business Week on January 13, 1975, declined to rule out the use of force in the Mideast.[16]

The American plan is drawn up, it is said, in such a way that in the event of a fifth Arab-Israeli war and another oil embargo, one marine division of about 14,000 men would be sent to the Persian Gulf. At the same time, a second task force consisting of nine infantry battalions of the 82nd Airborne Division, now based at Fort Bragg, North Carolina, would be flown to the Israeli airbase in the northern part of Negev. From the Israeli airbase American C-5A and C-141 jetcraft would carry paratroopers, accompanied by an escort of Phantom fighters, to complete the final 1,000-mile journey, arriving at Dhahran airfield at least three days before the seaborne marines. The job of the paratroopers would then be to seize Dhahran airfield, evacuate American personnel from the area, and call in reinforcements to capture oil tankers at Ras Tanura. Once this has been accomplished the paratroopers would move inland to occupy the largest oil field at Ghawar, only 70 miles from Dhahran. The marines' job would be to complete the occupation of Ghawar and nearby Abqaiq oil field, and send patrols north toward Kuwait and toward controlling the Strait of Hormuz on the Persian Gulf. The strait is most important as around 23 million barrels of oil a day, about one-half of the non-Communist world's current import, are carried by tankers that pass through the shallow 26-mile-wide strait. It has come to occupy a position as strategic in world politics as that of the Strait of Gibraltar or the Panama Canal in the past. At the same time using helicopters and boats, American troops could occupy smaller oil fields near Doha, Adma, and Dubai. The air cover against the Saudi air force would be provided by the marine division's own units consisting of eight Phantom squadrons, two reconnaissance squadrons, and another eight squadrons of light and medium bombers. In this operation there would be some 40, 000 U.S. forces involved in the Middle East. The small Saudi army, consisting of about 36, 000 troops, is deficient in training and equipment and could not be regarded as a match for the American expeditionary forces. If the Saudis are prudent they would, in such an event, avoid combat completely and make concessions without battle.

The 1975 understanding between the United States and the Sultan of Oman reportedly includes the facilities of a military base on Masirah Island at the gulf's southern end, which is suitable for stationing U.S. submarines and heavy aircraft. Added to this is a semiprivate American army, numbering 2, 200 engaged in training the Saudi army, and possibly there is an equal number of army personnel stationed in other Arab countries.[17] Iran has 6, 500 American military advisers, helicopter pilots, army, and civilian technicians,[18] many of whom could be moved to the Arab side of the gulf to assist in the American war efforts, if necessary. These semiprivate American armies would undoubtedly attempt to prevent the burning of the oil fields which the Arabs, it has been said, intend to do in case of foreign military invasion. (The flames of burning oil fields could be stopped quickly by sea-lifted Texas oil field fire-fighting equipment. The U.S. Army Engineer Corps is capable of repairing damages to ravaged oil fields in two months' time.) The private American army could, together with their trainees and friends in the local army, sabotage Arab resistance plans in other ways by collaborating with the invaders. Again it is through this help that American marines could easily land at Dhahran. Following an invasion, the local army could be installed as a friendly or puppet regime to run the government. The prospect of coming to power through American assistance would offer enough inducement to the Arab armies to cooperate with the American military mission of adventure in the Mideast.

Israel, which joined Britain and France in the Suez War in 1956, might also undertake preemptive raids against Saudi Arabia, Kuwait, and the UAE, with a view to destroying the financial bases of the Arab military might. Since 1973 these Arab nations have been the main sources of finance for the confrontation Arab states' rearmament programs; therefore, Israel's next attempt could very well be to cause as much damage as possible by attacking those oil-rich Arab countries. Incidentally, in all past wars Israel has opened up new battle fronts each time; so it is reasonable to say that in the next Mideast war the oil-rich Arabs may be Israel's new targets. While this is possible, American seizure of the Arab oil field in a blitzkrieg attack is probable.

More important, apart from the political effect of such a step on international relations, is possible military response by the Soviet

Union. The Soviet Union would likely respond militarily, not in a competitive manner but rather in cooperation with the United States, though in a different direction which would be the seizure of Iran's oil.

Over the years since World War II, U.S.-USSR cooperative relationships have been based on American assertiveness in international conflict situations, such as in Berlin, in Cuba, and in the Middle East itself. U.S. assertiveness is based on the relative military and economic strength of the two superpowers. While there is no controversy as to American economic superiority, some analysts such as George Ball believe that the Soviets have reached nuclear parity with the United States.[19] But there is likely to be some kind of a general understanding between the two superpowers not to engage in nuclear confrontation. In comparison to the energy crisis, the Arab-Israeli dispute is only a marginal concern to the superpowers. Therefore, the sharing of Middle East oil resources could be a major test for detente.

Barring the possibility of nuclear confrontation between coequal powers, the Soviet military threat to American maneuvers in the Middle East may be considered more than manageable. In such an eventuality, all or almost all of the NATO allies of the United States would certainly support American military overtures in the Middle East for the purpose of securing oil for energy-thirsty Western Europe. Despite the impressive Soviet buildup in the Mediterranean, its squadron and supporting land-based craft are confronted with several weaknesses and handicaps:

> Most important, it is faced with the overwhelming naval and tactical air power of NATO. First, naval forces assigned to Allied Forces South retain a superiority in conventional surface fire-power, amphibious landing capabilities, naval air support, reconnaissance, fleet escort, and submarine resources. . . . This superiority would allow NATO sufficient forces to seal and quadrant the Mediterranean, while leaving the Sixth Fleet, the British naval commando assault group, plus other NATO contingents, free for offensive operations.[20]

A nonmilitary restraint on Soviet initiative in the Middle East is its failure to protect its proclaimed interests in crises:

> Indeed, when potential confrontations with the West have developed, as in the 1967 June War, the November 1968 Jordanian crisis, the 1969 Lebanese crisis, and the 1970 Jordanian civil war, the Soviets were extremely careful to insure that their forces did not appear in a provocative posture.[21]

Even in the October war the Soviet Union finally gave up its announced policy of sending volunteers to help protect Arab interests. The Russians, however, have been reckless in their political dealings with the Americans, but only after the crises have peaked or have passed.

In the perilous scenario of the forthcoming American military invasion of the Middle East, the Russians are likely to behave with more prudence since their own oil interests are at stake. There is also a long history of Soviet interests in the warm water ports of the Mediterranean. Oftentimes it has succeeded in establishing control over these ports but only in cooperation with the West. It failed to maintain these controls when its interest was in conflict with the West. Soviet foreign policy, like that of any other nation, is determined by its perpetual interest and not by friendship with nations, which is short-lived. Russia had a heyday in its relationship with Egypt for some time. Now it is cultivating Iraq, an ideologically Baathist nation, not Communist. While continuing to cultivate Iraq, Syria, Egypt, and the People's Yemen and to supply the Arabs with modern military hardware so as to enable them to fight their war with Israel, the USSR could possibly acquire control of Iranian oil.

Of all the interests the superpowers have in the Mideast, oil is most vital for their progress. This is of major importance for the Americans and their Western allies in the immediate future until oil from Alaska and the North Sea is available for consumption, which will take a minimum of at least two and three years respectively. The oil shortage in the Soviet Union is becoming more and more serious and, by all indications, will take a turn for the worse by 1980. Since the prospect for permanent settlement of the Arab-Israeli conflict is waning, there could develop a cooperative effort between the superpowers to acquire oil by force from the Middle East, as both need a secure oil supply.

After the Vietnam experience, Americans are not willing to spill their blood for other people's wars, but it would be a lot easier to engage in a Mideast war, not because of American Jewish support but because it would be directed toward the prevention of the use of oil for destruction of the financial and economic system of the Western world. "The U.S. cannot again permit oil to be used as an essentially military weapon without a suitable military response."[22]

A cursory glance at the modern history of Russo-American relations would show that when both were faced with a common enemy they joined hands and fought side by side. Whenever the United States has tended to become assertive the Soviet Union has become either cooperative or, at least, inactive. Judging from this historical perspective it may be safe to say that the Russians would probably cooperate with the Americans against a common enemy—in this case the "enemy" is the oil shortage. The time is fast approaching when detente might prove useful in obtaining oil.

Because of their existing sociopolitical differences, it is impossible to predict accurately the future course of Soviet-American relations, but an encouraging sign is that both sides have joined together in a number of interests, such as trade from time to time and cultural contacts, which are indicative of a slow but gradual movement toward detente:

> Underlying the Soviet-American detente is the conviction that whatever the difference between the superpowers an easing of tension is preferable to an indefinite continuation of a dangerous and essentially futile confrontation. In one sense, the new Soviet-American relationship will depend on whether the political will toward detente outweigh the mutual hostility that remains and the desire of both not to be outdone militarily, diplomatically, or economically.[23]

THE INTERNATIONAL ENERGY AGENCY

It was to cope with OPEC's monopolistic control of oil that, at the height of the energy crisis in February 1974, U.S. Secretary of State Kissinger first called for joint action by consumer nations. Thirteen foreign ministers of the major oil-consuming nations gathered in Washington on February 11 for a three-day conference to discuss an energy policy. The nations represented at the Washington energy conference were Belgium, Canada, Denmark, France, the Federal Republic of Germany, Ireland, Italy, Japan, Luxembourg, the Netherlands, Norway, the United Kingdom, and the United States. The conference discusses a seven-point program for cooperative action presented by Kissinger and other proposals for an energy program. In effect the communique issued at the end of the conference incorporated much of the Kissinger proposal: (1) the cooperation on conservation of energy; (2) the development of alternate energy sources; (3) research and development; (4) the sharing of oil in emergencies; (5) international financial cooperation; (6) an invitation to less-developed countries (LDC) to join the major consumers of oil; and (7) cooperation between consumers and producers.[24]

The United States, not being as dependent on Arab oil as some of the others, could follow an independent or pro-Israeli policy in the Mideast. On the other side of the Atlantic, America's allies were being humiliated by having to assume, under pressure, a pro-Arab policy in order to keep their oil supplies coming. The disarray in the NATO alliance was evident when at the height of the fourth Arab-Israeli war Secretary Kissinger launched his shuttle diplomacy from Washington to Moscow and the Mideastern capitals without consulting U.S. allies in Western Europe. The Atlantic allies refused to permit U.S. aid to Israel from bases in their territories. This disarray in the NATO rank and file encouraged the Arabs to believe that an oil embargo against the nations supporting Israel would produce public and diplomatic opinion against Israel, as indeed it did in Europe. The Kissinger strategy in convening the conference, aside from energy-sharing proposals, had its political significance in renewing the confidence in the NATO military alliance.

As expected, the oil producers vigorously objected to the convening of the conference. The Arab oil producers, particularly, were disturbed as far-reaching cooperative efforts were being made to create an international energy agency to deal with such emergencies as embargoes by sharing available oil and by cutting consumption and using stocks on an equitable basis. The secretary of state's global energy program elicited muted response from most Arab capitals.[25] Nevertheless, the conference marked the beginning of the renewal of confidence in the Atlantic alliance under American leadership.

The conference agreed to establish a coordinating group to direct and coordinate actions. France dissented from this and some other declarations of the conference. France saw within the American-initiated conference of major oil-consuming nations a device by which to reassert its economic domination of Europe:

> French Foreign Minister Michel Jobert tried to sabotage the conference. He failed when the delegates voted 12 to 1 to cooperate with the United States on oil policy. But France soon appeared to hold the upper hand after all. The Common Market countries acquiesced in a French move on March 4 to authorize economic and technical cooperation with 20 Arab states independently of the United States. This action was what prompted President Nixon's outburst in Chicago on March 15. He warned that economic "hostility" in Europe would bolster arguments in Congress for withdrawing American troops from the continent.[26]

Addressing the U.N. General Assembly on September 18, 1974, President Ford sought new approaches to international cooperation to overcome the problems that the world was facing as a result of the oil crisis.[27] Then the oil producers got a double warning from President Ford and from Secretary Kissinger that the global economic chaos caused by the high prices of oil threatened to engulf the world into a general depression. Speaking at the World Energy Conference in Detroit with representatives of 69 nations present, Ford said that no nation can benefit at the expense of others "except for the very short term and at very great risk."[28] He said that just as Americans were challenged by Project Independence, the world faced a related challenge that required a Project Interdependence. Convinced that an oil crisis can be avoided through concerted efforts by the industrial powers, the United States shelved its Project Independence for the time being and began to work for the Project Interdependence. To forge unity among the oil-consuming nations of Europe on the concept of Project Interdependence, on the same day (September 23, 1974) that President Ford was addressing the energy conference in Detroit Secretary Kissinger told the U.N. General Assembly that the nations of the world have been using "old patterns of thought and action"[29] to meet new political and economic problems—the obvious reference being that the oil producers were taking revenge on Europe for its past colonialization of the world when some resources were transferred from today's Third and Fourth Worlds to Europe. Europe is becoming impoverished because its money is now being transferred to countries that export oil to the continent.

Meanwhile, the United States and other industrial countries met in Brussels to give concrete shape to the proposed international energy agency. The Kissinger plan was still undergoing modifications. Ultimately what emerged was a four-point program of international cooperation that Kissinger hoped would help avert disruptions of the Western and Japanese economies. Finally, the United States, Canada, Japan, Turkey, and 12 European countries formally created the International Energy Agency (IEA) in Paris on November 15, 1974. The four main features of the IEA, or counteroil cartel as its critics call it, are (1) the establishment of an energy-sharing agreement among the major consumers in case of a new embargo; (2) the setting up of cooperative conservation and energy-development programs; (3) the establishment of a $25 billion fund to recycle petromoney into deficit countries; and (4) the convening of a conference among producer and consumer countries.[30]

France abstained from voting for the IEA, which came into being under the Organization for Economic Cooperation and Development (OECD). While Greece and Finland also abstained, the five other members of the 24-nation OECD voted in favor of the creation of the IEA but did not join it.[31] The IEA has an executive committee comprising all the member countries. The committee's decisions are binding upon all the member countries, except for contrary arrangements. Decisions require a 60 percent vote, which means 89 votes out of a total of 148.[32] Votes are weighted by population with the United States having 51; Japan, 18; and West Germany, 11. In this weighted voting the United States has a few more than one-third of the total vote.

The energy-sharing, stockpiling, and conservation programs are already a part of IEA. The creation of a recycling fund has been organized through the IMF. Some of the Arab nations, such as Kuwait in its attempt to diversify foreign investment programs, decided to finance an oil pipeline from Yugoslavia's Adriatic Coast into the Soviet bloc's oil distribution system. The task of holding a tripartite conference between the oil producers, industrial oil consumers, and the poor nations was made easier when a Franco-American summit conference on December 16, 1974, reached a compromise over their differences in dealing with oil-exporting nations. It was in a spirit of harmony and reconciliation that the United States agreed to cooperate with France which has not joined the 16-nation IEA. The French fear of American domination seems to have been reduced by the fear of the common enemy—the oil crisis. However, the tripartite conference, convened in Paris, broke down in complete disagreement.

Cooperation for mutual advantage among oil consumers obviously implies confrontation with the producers. The United States, the creator of IEA, has already confronted OPEC rhetorically. Proud of the seemingly impossible task of uniting the industrial oil consumers the American secretary of state said dramatically that the United States "will never permit itself to be held hostage—politically or economically."[33] While the major oil consumers were demonstrating their unity in evolving a common emergency program of oil sharing, OPEC on its part in further demonstration of its unity, increased oil prices by 38 cents a barrel within a short time of the formation of the counteroil cartel by the consumers. Reportedly irritated soon after the formation of IEA by a news photo of British Prime Minister Wilson embracing former Israeli Premier Golda Meir during her visit to London, Saudi Arabia—the largest OPEC producer—ordered that sterling would not be accepted in payment for oil, which drove the pound down to its lowest price in history.[34]

The battle for maintaining or even raising the already mounting prices of oil and the counterbattle to lower them are inexorably moving the oil producers and the consumers toward confrontation. Now the question is what the industrial nations do about OPEC's effort to invalidate the forces of the free market economy. Contrary to the hopes that Saudi Arabia would use its influence to bring oil prices down, it joined with others to increase the prices. The issue, then, is not a financial crisis but a question of prosperity of the industrialized world.

It is for this reason of prosperity that Kissinger described military action in the Middle East to lower oil prices as "a very dangerous course." But he left open the possible use of force to prevent "strangulation of the industrialized world." He expressed these views in an unusually blunt year-end interview with Business Week magazine.[35] When asked if he considered military action on oil, Kissinger responded:

> A very dangerous course. We should have learned from Vietnam that it is easier to get into a war than to get out of it. I am not saying that there's no circumstance where we would not use force. But it is one thing to use it in the case of a dispute over price, it's another where there's some actual strangulation of the industrialized world.[36]

The United States is taking its IEA partners into confidence and slowly mutual linkages between their economic and energy policies are developing. The Paris conference of major oil consumers, oil exporters, and poor countries in April 1975 failed as there was no common political and economic base for a worldwide indexing of oil price with other commodities and manufactured goods; there was more unity among the IEA group than among the oil-exporting and poor countries. However, if the situation in the Mideast deteriorates and the threat of war accompanied by an oil embargo becomes real, IEA members are likely to evolve a mutually beneficial war policy and invade the oil installations on the Persian Gulf even without Soviet cooperation, which could be extremely risky and only used as a last resort to prevent dissolution of Western economic structure.

CIVILIAN CONCEPT OF MUTUAL DETERRENCE

To the students of international relations the concept of deterrence means

> activities undertaken by a state or group of states to discourage other states from pursuing policies unwanted by the deterring state or states. Deterrence involves a strategy of threatened punishment or denial to convince others that the costs of their anticipated action will outweigh the gains. The means by which states pursue policies of deterrence include increasing their general military capabilities . . . and threatening reprisals.[37]

This concept has worked to maintain the present-day international system since the development of parity or near parity in nuclear striking power by the two superpowers, as neither could attack the other without mutually destroying each other. Robert S. McNamara, a former U.S. secretary of defense, writes in his book *The Essence of Security: Reflections in Office* that because of the assured destructive capability of the two superpowers a nuclear holocaust in the world has been avoided.[38]

From the strategic concept of deterrence we now turn to the non-military concept of deterrence. This is a concept which can be used to protect the vital international interests in the production and distribution of the most basic resources of the world, particularly food and oil—the two most important resources for the peacetime survival of humankind. Food for oil, that is, no oil, no food, is the basis of this concept. It would seem unethical to think of food for oil, but as will be seen below, both have been used in the world for political gains. Arguments about fairness are expedient and cut both ways.

Modern man seldom, if ever, eats food produced by solar energy; now he mostly eats food produced by oil.[39] The most basic human need is food. Its production is threatened when there is a shortage of oil supply due to its embargo or production cut. Almost all nations endowed with surplus oil are chronically deficient in food production. On the other hand, nations which produce surplus foodgrains do not all have surplus oil supplies. It is, therefore, in the vital interests of the oil producers to maintain the supply of fuel to the food producers so that food production is not hampered by shortage of oil. Despite this, if oil producers embargo the oil supply or cut down on production for political reasons, the suitable political response is to embargo the supply of food to the oil exporters.

The major world food suppliers are the United States, Canada, Australia, and France. The major oil exporters are the 13 OPEC countries. The nations of Latin America, Eastern Europe, including the

MAP 5

The Oil Fields of Eastern Saudi Arabia

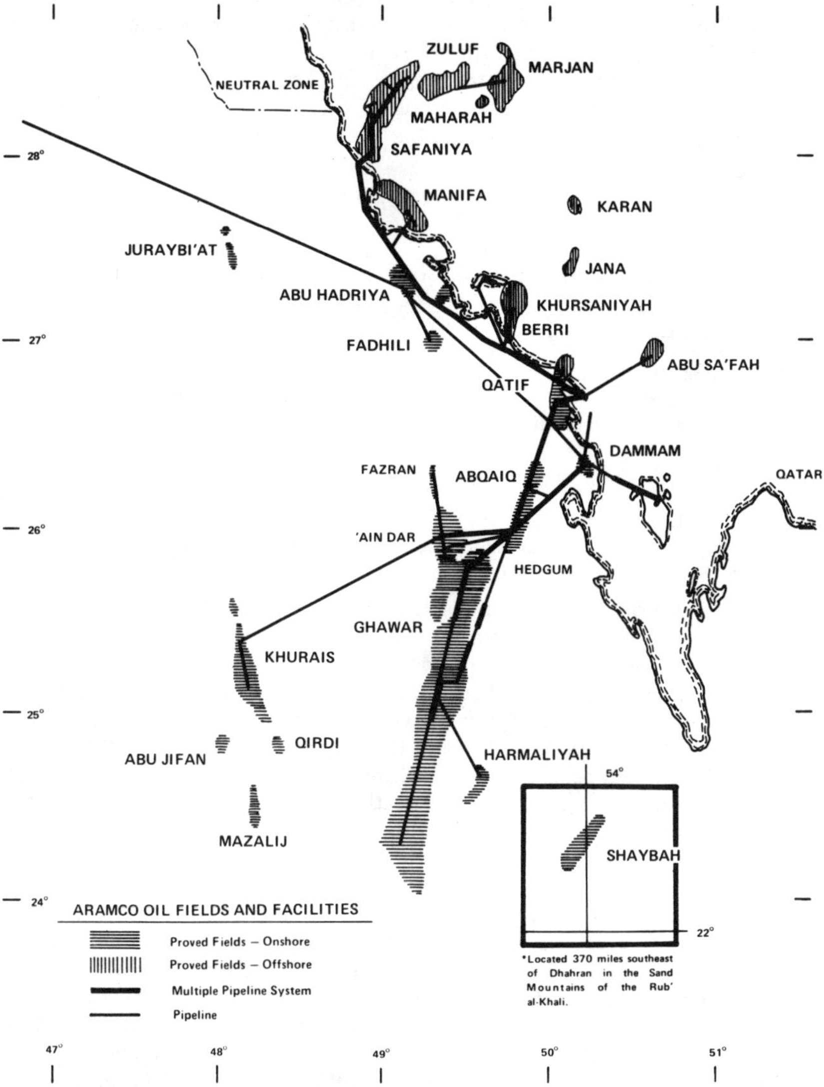

Source: Aramco. 1973. Annual Report for 1973, p. 9, reprinted in Ramon Knauerhase, The Saudi Arabian Economy (New York: Praeger, 1975), p. 9.

Soviet Union, and South Asia were for many years exporters of food to the world. The explosive population growth and climatic influence on food production have changed these and many other nations in the world from exporters to grain-importing countries. North America and Australia are the two geographic regions now producing exportable food surpluses. The United States is the world's leading exporter of wheat, feedgrains, rice and soybeans. The United States and Canada control a greater share of the exportable grain supplies of the world than the whole Middle East region does of oil.[40] Table 8 shows the disparity in grain trade in the various regions of the world; plus it indicates net exports and minus net imports.

TABLE 8

Regional Disparities in Grain Trade, Selected Years, 1934-73
(million metric tons)

Region	1934-38	1948-52	1960	1966	1973*
North America	+ 5	+23	+39	+59	+88
Latin America	+ 9	+ 1	0	+ 5	- 4
Western Europe	-24	-22	-25	-27	-21
Eastern Europe and USSR	+ 5	—	0	- 4	-27
Africa	+ 1	0	- 2	- 7	- 4
Asia	+ 2	- 6	-17	-34	-39
Australia	+ 3	+ 3	+ 6	+ 8	+ 7

*Preliminary fiscal year.

Source: Cited in Lester R. Brown, In the Human Interest (New York: W. W. Norton & Co., 1974), p. 81.

No region or continent of the world is endowed with all the basic resources needed to nourish modern civilization. The North American continent (the United States and Canada) is presently blessed with a monopoly of exportable foodgrains. On the contrary, the Middle Eastern Arab countries are not blessed with a monopoly of oil. As such there does not exist a parity between the supply of the two most vital commodities in the two regions. The balance is, therefore, heavily weighted in favor of the United States. But without affordable oil, farming in the United States may be affected drastically, not to speak of climatic and other conditions. Indications show that the need for American food aid and exports to the world will be much greater in the decade 1975-85 than in the previous ten years.[41] With such a situation the never-ending spiral in the price of oil or another embargo of it in the event of a Middle East war is self-defeating. This is so not because of the prospect of diminishing crop yields in the United States due to an inadequate oil supply, but because of the hope of the populous Third and Fourth Worlds for the "Green Revolution" has been crushed—not so much by its failure but by the failure to evolve an effective control of population which is growing faster than food production.

A majority of the world's 160 nation-states depend on food supplies from the United States alone. A majority of the nations also depend on outside oil supplies, not on one nation but on a group of nations which are at the moment working as one nation in collaboration of their oil policies. If OPEC unity continues to maintain the price hike and OAPEC invokes another oil embargo, then the United States may be forced to resort to an embargo on the embargoers by cutting its monopolistic supply and sale of food to the hungry world, particularly to the Arab world's oil producers.

The Arab oil embargo of 1973 has left a lasting impact on the energy-intensive agriculture in the United States. American food aid to the world is as politicized as the oil embargoes. The U.S. food aid to the world began immediately following World War I, when food was shipped to Europe to prevent hunger there. After World War II massive food aid was again given to Europe, ending in the late 1940s. Since then the U.S. food aid program has been diverted to the Third and Fourth Worlds. Between 1965 and 1972 America provided 84 percent of all the food aid to these nations. The nations of the Second World have become major grain importers from the United States. In the future the major powers would be at least partially dependent on food imports from the United States. Even taking into consideration optimistic projections about food productions in the Third and Fourth Worlds almost all members of OPEC and LDCs will remain dependent on the United States for food supplies during the decade 1975-85. This could give the United States a measure of political and economic power far greater than it has had.

The politics of American food aid were amply demonstrated at the World Food Conference in Rome in November 1974. Throughout the conference many people were dismayed at the contrast between the generous keynote address of U.S. Secretary of State Kissinger and the tight-fisted stands taken by the American delegation leader, Agriculture Secretary Earl Butz. In announcing at a news conference in Rome that President Ford has refused to permit the American delegation to commit a million-ton increase in emergency food aid to nations threatened with famine, Secretary Butz admitted later that food is a mere "tool in the kit of American diplomacy."[42]

Throughout the conference the traditional food donors, such as the United States and Canada, made it clear that an annual target of 10 million tons over the next three years for all international food aid could only be reached if oil-exporting countries contributed by financing food imports of MSA nations. American people in general are humanitarian. Their humanitarian concern was expressed in an editorial by the New York *Times* which said:

> For some Americans, pictures of hungry children or the knowledge that 32 countries are endangered by the food crisis will be sufficient reason to support whatever policy changes or whatever sacrifices may be called for.[43]

But public sympathy is not enough when legally sovereign nations confront each other; there are always huge gaps between interest articulation and interest aggregation. The MSAs see little difference in a threefold increase in the price of American foodgrains and a fourfold increase in the price of oil. They are unable to buy both commodities without foreign assistance. But there is a difference. The MSA nations are critical of the U.S. food program in the world but have nothing to say about the embargo of oil and its price escalation. In the international forums they side with petropowers without even thinking for a moment that the oil politics drove up the cost of production of fertilizer, planting, harvesting, and transporting grains. Piqued by these developments, Butz indicated that the United States no longer felt compelled to feed the MSAs.[44]

Two more examples will suffice to show the extent of American food aid politics. Bangladesh, probably the most severely affected of the 32 MSA countries, was told by the United States in late summer of 1974 that unless it stopped selling its gunnysacks to Cuba it would be ineligible to receive American loans on easy terms to buy food.[45] This was done shortly before President Ford waived restrictions, resulting from Egypt's trade with Cuba. The Ford administration authorized loans to Cairo totaling $27 million for the purchase of U.S. foodgrains.[46] Another MSA country is India which is an unofficial "opposition member" of the nuclear club but deficient in food production. According to Daniel P. Moynihan, former American ambassador to India, the U.S. Congress imposed restrictions on India's receiving 300,000 tons of foodgrains on a low-interest loan. Moynihan said that "he sees the day when countries might refuse American aid because of restrictions imposed by Congress."[47]

These are typical examples of politics in the food aid program. Let us now review the contention of the oil exporters that because of the increase in the prices of food and other essentials they are forced to raise the prices of oil. U.S. wheat prices jumped 192 percent in the two years prior to November 1974.[48] Arab oil prices rose sharply by 400 percent in little over one year's time before November 1974. The rise in American wheat and other foodgrain prices is due mainly to the free market surge caused by disastrous crop failures in the Soviet Union, China, and around the globe. Added to this are the costs of land, labor, and production in the United States which are the highest in the world. The justification for oil price increases is that the industrial countries exploited oil exporters for too long and the terms of trade were unfavorable for them.

The energy crisis and the food crisis are inextricably linked. Nevertheless, there is a difference in determining the prices of oil and food. Oil involves international power politics, seldom guided by democratic considerations. All OPEC governments are nondemocratic and oil resources are owned and sold by these governments without due regard to the economic law of supply and demand. Prices are raised by imposing embargoes and other conditions prejudicial to free market economy. On the other hand, American lands are mostly owned by millions of individual farmers, and the production and sale of food dominate regional and national politics in the United States. There is, however, no control of the prices of food by the governmental authority as national and international buyers are allowed to purchase foodstuffs through open tenders and competitive bids in the United States.

Even in 1975 production of a barrel of oil in Saudi Arabia costs 12 cents and is slightly higher in other Arab countries and Iran. The OPEC nations are charging $10.12 for a barrel of oil. The consumers feel that price of oil should bear some relationship to the cost of production. During and following the Arab oil embargo the multinational oil companies earned fabulous sums of money as profits. Currently, however, they are only making 20 cents to 50 cents per barrel.[49] Thus, Saudi Arabia makes a net profit of at least $9.50 per barrel of oil sold. Neither Saudi Arabia, nor Iran, nor other exporters need that much profit to compensate for inflation and to pay the higher prices for other commodities, which are relatively less costly than oil in every case. On this subject, a leading oil consultant quotes the U.N. index of world export prices of manufactured goods of 11 industrial countries and the index of the Saudi Arabian government revenue until 1973 and gives his own estimates for 1974 as shown in Table 9.

TABLE 9

World Export Prices Versus Saudi Oil Revenue, 1955-74
(base: 1959=100)

Year	Index of World Export Prices	Index of Saudi Oil Revenue per Barrel
1955	93	108
1960	102	99
1965	107	110
1970	123	116
1971	129	166
1972	140	190
1973	164	508
1974	184	1,224

Source: John H. Lichtblau, "Arab Oil and a Settlement of the Middle East Conflict," in *After the Settlement: New Direction and New Relationships* (Papers delivered at the 28th Annual Conference of the Middle East Institute, Washington, D.C., October 12, 1974), p. 9.

As Table 9 shows, the OPEC nations' argument that the giant increase in oil prices is necessary to offset world inflation had some validity between 1955 and 1970. From 1970 to 1974, world export prices of manufactured goods have risen about 50 percent and Saudi government oil income per barrel has risen 950 percent.[50]

Another argument is that in the past the industrial countries exploited the oil exporters. This was true until 1950 as far as Saudi Arabia is concerned. Since then, because of the establishment of the 50-50 profit-sharing formula, it is difficult for Saudi Arabia to make a case for exploitation. The position of other oil producers in the Middle East is not much different as most of them followed the Saudi example of an equal profit-sharing formula. Furthermore, a case for exploitation is difficult to make inasmuch as the oil companies made most of their profits at the crude oil producing level and not in downstream operations.[51] The profit margin for the companies was further reduced by a subsequent introduction of a 60-40 split. In 1974, under difficult operating conditions and constant pressures, Aramco agreed to sell their remaining 40 percent ownership to Saudi Arabia. The Saudi government will pay $2 billion for almost all Aramco facilities. This cost is estimated to be less than one month of Saudi oil earnings.[52] Inevitably the Saudi takeover will be followed by the rest of the oil producers in the region.

Reverting to our discussion of food for oil, it can be said that from the beginning of the American food aid programs they were a humanitarian gesture for the disposal of vast surplus crops. The surpluses are now disappearing as a result of large sales abroad and a drop in production. The selling of American food means starvation to some of the countries of South Asia, Southeast Asia, and the Sahel area of Africa, which have very little means to buy food. Although the United States has made attempts since 1973 to rebuild the depleted reserves of food by lifting acreage restrictions on corn, wheat, and sorghum, the food stocks are no longer large and easily available. It is not surprising, therefore, that the United States looks on food assistance more in terms of a politicodiplomatic advantage than in an economoral sense. It is even unable to maintain its superpower image of food donor as economic considerations at home outweigh moral considerations abroad. Due to these circumstances an actual embargo of food supplies to Arab nations during a war in the Middle East cannot be ruled out. In the past the United States embargoed food supplies to Egypt. Thus an automatic Arab oil embargo against the United States "is no longer the trump card it once was thought to be."[53]

Clearly there is one solution to these world problems—cooperation among interdependent nations. But resurging nationalism is causing nations to confront each other, and global cooperation is proving to be the most difficult of all problems. The civilian deterrence to further disruption in the free world economy is to make it abundantly clear to the oil exporters that, if the prices of oil do not reflect the free market economy, food supplies to those nations may be denied. This undoubtedly sounds cruel, but it is surely no less cruel to stop supplies of oil to the food producers. Ethical and moral considerations seldom, if ever, guide international politics, and the American food supply to the world has not been devoid of them. The key to deterrence is found in each side's hardened position. To be effective, a deterrence threat of food for oil should be fully credible to the oil exporters. Hopefully, this deterrence will avoid superpower confrontation and the Arab-Israeli conflict may be separated from the energy crisis in the world; thus having a better chance of settlement through peaceful means.

The deterrence concept does not imply actual application of food embargoes during peacetime, but a real threat of it is likely to forge co-operation between oil exporters and food exporters. As the fear of annihilation of the world has prevented the two superpowers from engaging themselves in nuclear confrontation, likewise the fear of starvation may prevent the petrocrats from devising ways and means of continuously increasing oil prices or imposing embargoes on the supply in case of another war in the Middle East.

In view of the end of the Western and Japanese nonpolicy on an oil program, the moves taken by IEA to accomplish stockpiling and the sharing of oil in case of another embargo, and the fact that OPEC is determined to maintain its unity, neither oil nor food can be denied selectively because of the fear of leakage. To make any potential embargo on the export of food effective, the embargo may have to be directed against all OPEC, especially OAPEC countries—moderates as well as radicals, friends and foes alike.

NOTES

1. Robert E. Hunter, "The Soviet Dilemma in the Middle East Part II: Oil and the Persian Gulf," Adelphi Papers (London: Institute for Strategic Studies, October 1969), p. 2.
2. John A. Berry, "Oil and Soviet Policy in the Middle East," The Middle East Journal, Spring 1972, p. 149.
3. Ibid., p. 150.
4. Ibid.
5. Hunter, loc. cit.
6. Christopher Tugendhat, Oil: The Biggest Business (New York: G. P. Putnam's Sons, 1968), p. 253.
7. Newsweek, December 30, 1974.
8. See the statement on the "tyranny of the majority" by the U.S. ambassador to the United Nations, John Scali, in Washington Post, December 7, 1974.
9. Richard H. Ullman, "After Rabat: Middle East Risks and American Roles," Foreign Affairs, January 1975, pp. 287-88.
10. Ibid., pp. 289-90.
11. Ball, op. cit., p. 8.
12. Newsweek, December 16, 1974.
13. Tareq Y. Ismael, The Middle East in World Politics: A Study in Contemporary International Relations (Syracuse: Syracuse University Press, 1974), p. 238.

14. Washington Post, September 24, 1974.

15. Miles Ingotus, "Seizing Arab Oil: A Blueprint for Fast and Effective Action," Harper's, March 1975. The article may have been inspired by the Jewish or other interested officials in the Pentagon to scare the Arabs. But the facts stated in the article seem to be accurate as several other sources also corroborate the same facts. As cited elsewhere, President Ford and Kissinger support seizure of Arab oil fields.

16. See Business Week, January 13, 1975.

17. Washington Post, February 20, 1974.

18. New York Times, February 20, 1975, and U.S. News & World Report, March 10, 1975.

19. Ball, op. cit., p. 8.

20. Lawrence L. Whetten, The Soviet Presence in the Eastern Mediterranean (New York: National Strategy Information Center, 1971), p. 35.

21. Ibid., pp. 40-41.

22. Joseph Alsop, "Perilous Scenario in the Mideast," Washington Post, November 6, 1974.

23. Editorial Research Reports on America's Changing Role (Washington, D.C.: Congressional Quarterly, 1974), p. 142.

24. See the full text of the communique in Washington Post, February 14, 1974.

25. Washington Post, November 16, 1974.

26. Editorial Research Reports on America's Changing Role, op. cit., p. 48.

27. New York Times, September 19, 1974.

28. New York Times, September 24, 1974.

29. Ibid.

30. Newsweek, December 23, 1974.

31. Organization for Economic Cooperation and Development, Press Release No. Press/A(74)48, November 15, 1974, and Washington Post, November 16, 1974.

32. Washington Post, November 16, 1974.

33. Time, November 25, 1974.

34. Newsweek, December 23, 1974.

35. Business Week, January 13, 1975, p. 69.

36. Ibid.

37. Jack C. Plano, and Roy Olton, The International Relations Dictionary (New York: Holt, Rinehart & Winston, 1969), p. 57.

38. Robert S. McNamara, The Essence of Security: Reflections in Office (New York: Harper & Row, 1968), pp. 51-67.

39. Paraphrased from Howard T. Odum, Environment, Power, and Society (New York: Wiley, 1971).

40. Lester R. Brown, In the Human Interest (New York: Norton & Co., 1974), pp. 81-82.

41. "To Nourish Humanity," Summary Proceedings of a Symposium on World Food and Population (Washington, D.C.: American University, 1974), p. 19.

42. Christian Science Monitor, November 18, 1974.
43. New York Times, November 19, 1974.
44. Editorial Research Reports on America's Changing Role, op. cit., p. 41
45. Washington Post, September 30, 1974.
46. Ibid.
47. Washington Post, December 30, 1974.
48. Time, January 6, 1975.
49. Ibid.
50. John H. Lichtblau, "Arab Oil and a Settlement of the Middle East Conflict," in After the Settlement: New Direction and New Relationships (Papers delivered at the 28th Annual Conference of the Middle East Institute, Washington, D.C., October 12, 1974), pp. 9-10.
51. Ibid., p. 10.
52. Time, January 6, 1975.
53. Newsweek, January 13, 1975.

CHAPTER

8

CONCLUSIONS

The abundant availability of energy resources, allied to technical feasibility and economic competitive compatibility, should in the long run provide sufficient energy to meet world needs. In the immediate future, that is, during the current decade, 1975-85, two nations will dominate the world energy scene. These nations are the United States and the Kingdom of Saudi Arabia. The United States will dominate the energy scene as the biggest consumer of oil, unless there is a change in American energy-consuming lifestyle; Saudi Arabia will dominate the world energy scene, since it holds the largest proven reserve of oil which is at present the most important primary source of energy.

The drive toward energy self-sufficiency in the United States through conservation, winning and recovery of oil from onshore and offshore areas, and various other measures as envisaged in the Project Independence will go a long way in reducing America's dangerous dependency on imported oil by 1980. However, in case of another Arab-Israeli war before 1980, the Arabs can be expected to use their oil weapon. In theory the Arabs could use their oil weapon again for military advantage at the battlefield and political benefit at home and abroad. But, in practice, another embargo will probably have more dangerous consequences for the Arabs than for the industrial consumers. While it is still too early to determine the full impact of Arab use of the oil weapon in the autumn and winter of 1973-74, it may be safely said that it forged cooperation among the industrial consumers. The formation of IEA and its oil-sharing policy in the event of another crisis shows that the major oil consumers are preparing in advance for any emergency situation. Another prolonged embargo of oil in the near future would, of course, be difficult for the industrial nations to bear. As a last resort, then, the industrial world headed by the United States would likely undertake an invasion of the oil fields in the Arab world. Most of the Western nations would prefer to follow the U.S. lead if it were made strong and credible for political as well as economic reasons. Already the United States has made its intention quite clear that, if pressed to the point of strangulation, it would not hesitate to undertake such an adventure. As noted previously, there remains the possibility of joint action by the industrial nations in the Arab world, while the Soviet Union concentrates on sharing the division of spoils in Iran. It would be much easier for the Russians to occupy Iran from the north. The Russians are not likely to join the war with the Arabs against the Americans, not only because of the existing detente between the two superpowers, but also because none of the three Arab nations—Saudi Arabia, Kuwait, and the United Arab Emirates—which may become vulnerable to American invasion are friends of the Soviets. Although the military balance weighs more in favor of the United States when taken into consideration with NATO capability, the proposed American strategy would be to avoid meticulously any military attack against radical Arab states friendly to the Russians. In this scenario the Soviets receive little or no provocation from the West to open a new theater of war such as in Berlin. Rather, the Russians will have nothing to lose, but they could probably regain their lost control of Iran which would offer them a secure energy source until oil from Siberia and other unexploited areas of the USSR becomes available for their consumption. The Soviets might even encourage such an American military expedition in the Mideast, as the war would hasten radicalization of the whole Arab world, which would be more advantageous for the Communists in the long run. The United States would be more concerned with stopping the immediate threat to Western strangulation, rather than preventing the growth of radicalism in the Arab world. The spread of radicalism or for that matter communism causes as much or more concern to the oil-rich Arab kingdom and sheikhdoms. Therefore, it might work as a limitation on the policy of petrolism to be followed again by the conservative Arab oil nations.

The essence of Soviet nonintervention in the Middle East is political as well as military. Once the Russians were satisfied that their political interests were served by not intervening in the Middle East they would not block American forces or warships deployed against the Saudis or the Kuwaitis. It is one thing for the Russians to open fire on Iranians, but it is quite a different issue to open fire on American troops. On this much higher level of military risk, the Russians would not be likely to escalate catastrophe in the Mideast. Soviet inaction would, of course, increase their political influence in the Arab world. If a question of life or death, however, the United States and the Western industrialized nations would like to have oil but not influence in the oil-producing nations.

A selective embargo against the United States may not stand a chance of success, since American multinational oil companies which continue to produce and market Arab oil could divert oil supplies meant for Europe to the United States, if the American government decides to protect the interest of oil companies. Any oil embargo to be effective must be a total and a much wider one; however, such a wider embargo is much too dangerous politically. A total embargo against the whole non-Communist industrialized world could be met not only by military response but by a counterdeterrence, that of a total food embargo by the main producers such as the United States, Canada, Australia, and the EEC countries. An energy doomsday, forecast by many pessimists, may not be far away, but it may be even closer for the oil exporters.

The Arab petrolism as a political weapon has almost no prospect in the long run, that is, after 1985 when the United States is likely to become an energy-autonomous nation again. Oil probably will be displaced as a primary commodity as substitution through nucosolar energy begins to take place in the United States as well as in Europe. In the short run the Arabs may attempt to use their oil weapon recurrently without being able to create an aggregate impact in the world. It is almost certain that the Arabs will use their new financial weapon in the event of a fifth war in the Arab-Israeli zone in the years immediately ahead. By withdrawing their petrodollars from the American, British, and other European banks, the Arabs could create widespread financial chaos and a serious threat to peace in the world. The recurrence of an oil embargo coupled with the use of a financial weapon may, however, entail dire consequences for Saudi Arabia and some of the other Arab oil exporters. In this regard the American military signals appear to have been received clearly by Saudi Arabia which refrained from making any serious comments.

As prudent and realistic as the Saudi rulers are they would not like to see such a situation develop. The Saudi rulers also know that the desert kingdom can only prosper through continued cooperation with the First World. Under these circumstances Saudi Arabia is not likely to participate in any future oil embargo, not to speak of leading it. If Saudi Arabia does not join prospective embargoes, their chances of success are almost nil. Consequently, there may not be any other Arab oil embargoes or the imposition of their financial squeeze in the future.

Turning to the other theme of our thesis that the Arab oil weapon has a limited scope in the settlement of the Arab-Israeli dispute, we should recapitulate here briefly the effect of the 1973 embargo on the peace initiative in the Middle East. How far did the oil embargo precipitate a peace settlement in the Middle East? Initially of course, while the war was still in progress, the embargo spurred the United States to work for a peace settlement. The Western European countries and Japan changed their pro-Israeli policy under the pressure of the Arab nations, becoming a little pro-Arab in their attitude toward the Arab-Israeli conflict. The united Arab action coupled with their improved performance on the warfield had its immediate effect. As assured to Saudi leaders, the United States moved in all seriousness to arrange a cease-fire and withdrawal of Israeli troops from the Suez Canal and part of the Golan Heights.

The Arabs hoped and the Israelis feared that the oil embargo would encourage a quick settlement of the politico-military crisis in the Mideast. Undoubtedly, some progress was made toward a disengagement and the speedy pullback of Israeli troops from some of the occupied areas, but the oil embargo quickly translated a regional conflict into global energy and financial crises. As the magnitude of these crises are felt more and more there is a corresponding indifference of the world toward the Mideast conflict, and the prospect of a peaceful settlement is waning. As the short-term hardship measures to stockpile and conserve energy and other devices to save gasoline and to increase its production are being implemented in the major oil-consuming nations, a psychic attitude of "tit for tat" is slowly developing in the industrialized world.

The progress in the territorial adjustment in the Mideast is not so much due to the impact of the oil embargo, although it could be argued at some length that way. The embargo was withdrawn (as of this writing) more than a year and a half ago and its impact is felt, not in the Mideast peace settlement but in the oil-consuming countries, more so in the oil-importing Fourth World. The economic consequences of the embargo and the quadrupling of oil prices, which have been noted previously, have shattered the development prospect of the poorest nations in the world. The ballooning price of oil has driven up the price of many necessities, for instance, fertilizer which has forced farmers in populous South Asian and many other countries to use far less fertilizer than normally. As a result food production has fallen off and hungry people in South Asia and elsewhere in the world are becoming hungrier and are beginning to depend more on Western food suppliers.

The United States, having suffered the least, although the oil embargo was mainly directed against it, has taken the initiative in the peace settlement, not because of the fear of another embargo as adequate measures are being taken to prevent such an eventuality in the future, but because of a desire to create a rapprochement between Israel and its Arab neighbors and to develop friendly relations with as many nations in the region as possible. Potentially, the most threatening trouble spot in the world is the Mideast where an arms race of unprecedented magnitude is going on, not only between the adversaries in the Arab-Israeli zone but among all the contenders in the region.

The fourth Arab-Israeli war was aimed not at the destruction of Israel but at regaining the occupied Arab territory. The Arab attitudes toward Israel have changed since the 1967 war. Until this time the Arabs demanded the complete dismantling of Israel. They modified their stand against Israel and facilitated the passing of U. N. Security Council Resolution 242 which called for an Israeli pullback from all territories occupied in 1967. The Arabs reaffirmed their stand not to "throw Israel into the sea" during the October 1973 war when a resolution for the imposition of an oil embargo was adopted in Kuwait. This Arab attitude of reconciliation with the Jewish state grew not because of love of Israel but because of their failure to regain territories by a war of attrition or diplomacy. The Arabs, by giving up their claim to Palestine, have endorsed a de facto existence of Israel. Perhaps this means that the Arabs would like the Palestinians themselves to decide the fate of their own homeland. The Palestinians, of course, have reaffirmed their claim to the whole territory of Palestine. Under such a situation, the Arabs could agree to certain territorial adjustment with Israel without waiting for the Palestinians to return to their homeland.

For a long time Arab oil was virtually immune from the effects of the Arab-Israeli conflict. King Faisal, though an avowed enemy of Israel, was loath to use the oil weapon against the United States, but was provoked by the spectacle of massive U.S. arms aid to Israel during the October war. This study leaves little room for Saudi Arabia to make use of its oil weapon again. The personal hope of King Faisal, a Muslim fundamentalist, to pray in the Al Aqsa Mosque in an independent Jerusalem ruled by the Arabs may not have materialized in his lifetime, even had the king, almost a septuagenarian and a sickly person, not been assassinated. Even though he was not able to visit an independent Jerusalem, King Faisal gained an image and position of power by virtue of the oil wealth bestowed on his kingdom. With more petrodollars pouring into the kingdom than it can spend in the foreseeable future, the Saudi power of oil-derived wealth is reaching around the globe. Saudi Arabia is increasingly being recognized as an oil superpower, and nations had been busy trying to appease King Faisal before his death. This was no mean achievement. Getting handsome doles of Saudi petrodollars, representatives of Muslim nations were crowding Riyadh for more blessings from the king. Gradually, King Faisal was assuming a leadership role in the Muslim world. Although both Saudi Arabia and Iran are spending billions of dollars in Western arms, Iran is more likely to emerge as a regional big power as it has the money and manpower needed to enable it to assume such a role of policemen, while the desert kingdom lacks manpower. All these are direct contributions of the oil wealth, an indirect effect of the 1973 oil embargo which quadrupled oil prices, importing billions of dollars to the hitherto impoverished oil-exporting nations. As detailed earlier, the oil embargo has had a limited impact on the peace settlement in the Mideast. The embargo's greatest effect is felt in the domain of the soaring accumulation of petrodollars, especially by Saudi Arabia. Two years have passed since the 1973 Mideast war and the Arab oil embargo, but there seems to be no end in sight to the deadlock in peace negotiation. It may be recalled that the oil embargo was not imposed for a cease-fire and subsequent disengagement agreements between Israel and the two confrontation Arab states; it was established specifically for securing an Israeli withdrawal within the territory it had before the six-day war of 1967. Had there been no oil embargo there would have been a cease-fire between the Arabs and the Israelis but no quick Israeli pullback from the Egyptian territory on the west side of the Suez Canal and the Syrian territory from the Golan Heights. Precisely, then, the Arab oil embargo accomplished only the initial Israeli pullback and exchange of prisoners of war.

In sum, the major finding from this study is that, although Saudi Arabia has almost one-fourth of the total non-Communist world reserve of oil, the desert kingdom alone is not capable of instituting an oil embargo against the industrialized world. Saudi Arabia may not participate in another oil embargo, if it is imposed by other Arab nations, because of the fear of invasion of its oil fields by the United States. In the absence of Saudi participation in any future embargo, oil cannot be used effectively as a weapon of diplomacy in any prospective Palestinian-Arab-Israeli war. It is more likely that oil will be displaced as a primary commodity and that it may be reduced to the position of a normal commodity in the world trade after 1985 when alternative energy development and domestic oil production may make the United States energy-autonomous and free it from the need to import oil. The Arab countries of the Mideast and North Africa have nearly 76 percent of the non-Communist world's oil reserves. In the immediate future this oil will continue to play its econopolitical role without having much success in crippling the industrialized world, but such oil politics may cause irreparable damage to the economic developments in the Fourth World. In default of a peace settlement in the Middle East, tension will continue in the region, but the waging of another major war by the disputants is unlikely in the near future. The "all-the-way" maximalist policy goal of the Palestinians to destroy Israel, a policy long abandoned by the Arab nations, in order to recreate their homeland may remain unfulfilled for a long time. Saudi Arabia has made use of its growing oil and financial power to exert diplomatic pressure against Israel. By using its oil weapon and petrodollar Saudi Arabia has managed on behalf of the Arabs and Islam to isolate Israel and Zionism internationally. Saudi Arabia will certainly assume a leadership role financially in the Arab world, as well as in the Muslim world. The Saudi kingdom is emerging as an oil superpower, a phenomenon indirectly contributed to by the 1973 oil embargo. Saudi Arabia, which represents a very tiny fraction of the world population and has developed so far little military potential, is emerging as a financial superpower in the world and the leading donor to the Muslim world. Still nearer home, the Saudi petrodollar is not only contributing to the rearmament and economic development programs of the Arab oil have-nots, but it is also greatly accelerating the process of modernity of tradition in the superoil kingdom. Saudi Arabia may never become a superpower in the same sense as we understand the term, but it is moving into a position where even the superpowers will have to listen when the Saudis speak. The reason is oil. Without oil Saudi Arabia probably would have remained a backward desert spotted with a few oases.

APPENDIX

RESOLUTION OF ARAB OIL MINISTERS

The oil ministers of the Organization of Arab Petroleum Exporting Countries (OAPEC) held a meeting in Kuwait on October 17, 1973, to consider the role of oil in the Arab people's current struggle to liberate their lands. Following through this question, the oil ministers,

Considering that the ultimate goal of the current struggle is the liberation of the Arab territories occupied by Israel in the 1967 war, and the restoration of the legitimate rights of the Palestinian people in accordance with United Nations' resolutions,

Considering that the United States is the principal and foremost source of Israeli power that enabled it to continue occupying their territories,

Considering that the industrial nations have a responsibility of implementing the United Nations Resolutions, and

Considering that the economic situation of many Arab oil producing countries does not justify raising oil production, although they are willing to make an increase to meet the demand of those industrial nations that are committed to cooperation in the task of liberating occupied territories,

Decided that each Arab oil exporting country immediately cut its oil production by a rate not less than 5% from the September production level, and further increases of 5% from each of the following months, until such a time as the international community compel Israel to relinquish occupied Arab lands, and to levels that will not undermine their economies or their national Arab obligations.

Countries supporting the Arab cause and those taking active and effective measures in compelling Israeli withdrawal shall not be affected by this production cut and shall continue to receive the same amount of oil supplies. The decrease in oil supply will be proportional to each outside country's degree of support and cooperation with the Israeli enemy.

The Oil Ministers also recommend that the United States should be subjected to the most severe cuts in the supply of crude oil and its refined products. This progressive reduction should lead to a total halt of all oil supplies to the United States from each individual country participating in the meeting.

Source: Embassy of Kuwait, Washington, D.C.

The Arab oil ministers met again in Kuwait on November 4, 1973, to discuss the question further and decided that initial production cut be 25% of the September level, and a further 5% from the production of each of the following months. The 25% cut should also include the complete halt of all oil shipments to both the United States and the Netherlands.

The Arab Oil Halt to the United States and Holland

The Arab oil producing countries have decided to halt their oil supplies to the United States and Holland and to any other country supporting Israel.

This decision is by no means directed against the peoples of the United States or Holland. It is in fact directed against their governments' hostile policies towards the Arab people.

The Arab people fully realize the interests of other people and want to develop closer ties with the people of the United States and Holland, who must also realize where their interests lie.

The Arab Oil Exporting Countries would like the American and Dutch people to know that the halt in oil supplies to their countries will continue until such a time as Israeli forces are fully withdrawn from all occupied Arab territories and the Arab people of Palestine regain their lawful rights.

The Arab Oil Ministers would like to draw the attention of the American people to the fact that the United States' Government itself adopted similar policies of banning shipments of arms, strategic material, such as oil and even food stuff to countries considered hostile to the United States.

BIBLIOGRAPHY

BOOKS

Abboushi, W. F. The Angry Arabs. Philadelphia: The Westminster Press, 1974.

Abir, Mordechai. Oil, Power and Politics: Conflict in Arabia, The Red Sea and the Gulf. London: Frank Cass, 1974.

Adelman, M. A. The World Petroleum Market. Baltimore: Johns Hopkins University Press, 1972.

Antonius, George. The Arab Awakening. Beirut: Khayats, 1955.

Arabian American Oil Co. The Arabia of Ibn Saud. New York, 1952.

__________. Middle East Oil Development. March 1956.

__________. Oil and the Middle East. Dhahran, 1968.

Assah, Ahmed. Miracle of the Desert Kingdom. London: Johnson, 1969.

Badeau, John S. The American Approach to the Arab World. New York: Harper and Row, 1968.

Barrows, Gordon H. The International Petroleum Industry. New York: International Petroleum Institute, 1965.

Black, C. E. The Dynamics of Modernization. New York: Harper and Row, 1967.

British Petroleum Co. Statistical Review of the World Oil Industry. London, 1956.

Brooks, Michael. Oil and Foreign Policy. London: Lawrence and Wishart, 1949.

Brown, Lester R. In the Human Interest: A Strategy to Stabilize World Population. New York: W. W. Norton & Co., 1974.

Brown, Seyom. New Forces in World Politics. Washington, D.C.: Brookings Institution, 1974.

Burrell, R. M., and Alvin J. Cottrell. Iran, the Arabian Peninsula, and the Indian Ocean. New York: National Strategy Information Center, 1972.

Byroade, Henry A. U. S. Foreign Policy in the Middle East. Department of State Publication, no. 4852, Washington, D. C., 1953.

Campbell, John C. Defense of the Middle East. New York: Frederick A. Praeger, 1960.

Caroe, Olaf K. Wells of Power. London: Macmillan & Co., 1951.

Chase Manhattan Bank. Capital Investment of the World Petroleum Industry. New York: Chase Manhattan, December 1974.

__________. The Profit Situation: A Special Petroleum Report. New York, April, 1974.

Cheney, Michael Sheldon. Big Oil Man from Arabia. New York: Ballantine Books, 1958.

Chomsky, Noam. Peace in the Middle East: Reflections on Justice and Nationhood. New York: Vintage Books, 1974.

Clark, James A. The Chronological History of the Petroleum and Natural Gas Industries. Houston: Clark Book Co., 1963.

Clark, Wilson. Energy for Survival. New York: Anchor Books, 1974.

Cleveland, Ray L. The Middle East and South Asia. Washington, D.C.: Stryker-Post Publications, 1973.

Commoner, Barry. The Closing Circle: Nature, Man & Technology. New York: Bantam Books, 1971.

Congressional Quarterly. Editorial Research Reports on America's Changing Role. Washington, D.C.: Congressional Quarterly, 1974.

__________. Energy Crisis in America. Washington, D.C.: Congressional Quarterly, 1973.

__________. The Middle East: U.S. Policy, Israel, Oil and the Arabs. Washington, D.C.: Congressional Quarterly, April 1974.

Darmstadter, Joel et al. Energy in the World Economy. Baltimore: Johns Hopkins Press, 1971.

Davis, David Howard. Energy Politics. New York: St. Martin's Press, 1974.

De Gaury, Gerald. Faisal, King of Saudi Arabia. London: Arthur Barker, 1966.

De Novo, John A. American Interests and Policies in the Middle East, 1900-1939. Minneapolis: University of Minnesota Press, 1963.

Ebel, Robert E. Communist Trade in Oil and Gas. New York: Praeger, 1960.

Eddy, William A. F. D. R. Meets Ibn Saud. New York: American Friends of the Middle East, 1954.

Energy Policy Project of the Ford Foundation. A Time to Choose: America's Energy Future. Cambridge, Mass.: Ballinger Publishing Co., 1974.

Erickson, Edward W., and Leonard Waverman, eds. The Energy Question: An International Failure of Policy. Vol. 1. Toronto: University of Toronto Press, 1974.

Evron, Yair. The Middle East: Nations, Superpower and Wars. New York: Praeger, 1973.

Falk, Richard A. This Endangered Planet: Prospects and Proposals for Human Survival. New York: Vintage Books, 1971.

Fanning, Leonard M. Foreign Oil and the Free World. New York: McGraw Hill, 1954.

Finnie, David H. Desert Enterprise: The Middle East Oil Industry in Its Local Environment. Cambridge, Mass.: Harvard University Press, 1958.

__________. Pioneers East: The Early American Experience in the Middle East. Cambridge, Mass.: Harvard University Press, 1958.

Fischer, Louis. Oil Imperialism: The International Struggle for Petroleum. New York: International Publishers, 1926.

Fisher, Carol, and Fred Krinsky. Middle East in Crisis. Syracuse: Syracuse University Press, 1959.

Fisher, John C. Energy Crisis in Perspective. New York: John Wiley and Sons, 1974.

Fisher, Sydney N. The Middle East: A History. New York: Alfred A. Knopf, 1959.

Frank, Helmut J. Crude Oil Prices in the Middle East: A Study in Oligopolistic Behavior. New York: Praeger, 1967.

Frankel, P. H. Essentials of Petroleum. London: Chapman and Hall, 1940.

________. Mattei: Oil and Power Politics. New York: Praeger, 1966

________. Oil: The Facts of Life. London: Weidenfeld & Nicolson, 1962.

Freeman, S. David. Energy: The New Era. New York: Vintage Books, 1974.

Gabbay, Rony E. A Political Study of the Arab-Jewish Conflict: The Arab Refugee Problem. Paris: Libraire Minard, 1959.

Glubb, John Bagot. The Middle East Crisis: A Personal Interpretation. London: Hodder & Stoughton, 1967.

Hammond, Allen L. et al. Energy and the Future. Washington, D.C.: American Association for the Advancement of Science, 1973.

Handelman, John R. et al. Introductory Case Studies for International Relations: Vietnam/TheMiddle East/The Environmental Crisis. Chicago: Rand McNally Co., 1974.

Harari, Maurice. Government and Politics of the Middle East. Englewood Cliffs, N.J.: Prentice-Hall, 1962.

Hartshorn, J. E. Politics and World Oil Economics. New York: Frederick A. Praeger, 1967.

Hazard, Harry W. Saudi Arabia. New Haven, Conn.: HRAF (Human Relations Area Files) Press, 1956.

Hill, P., and R. Vielvoye. Energy in Crisis, A Guide to World Oil Supply and Demand and Alternative Resources. London: Robert Yeatman, 1974.

________. Energy in Crisis. New York: Drake Publishers, 1975.

Hopwood, Derek. The Arabian Peninsula. London: Allen& Unwin, 1972.

Hoskins, Halford L. The Middle East: Problem Area in World Politics. New York: Macmillan Co., 1954.

Howard, Harry. U.S. Policy in the Near East 1954. Department of State Publication, no. 5801, Washington, D.C., May 1955.

Howe, James W. The U.S. and the Developing World: Agenda for Action 1974. New York: Praeger, 1974.

Howe, Irving, and Carl Gershman, eds. Israel, the Arabs & the Middle East. New York: Bantam Books, 1972

Howley, Dennis C. The United Nations and the Palestinians. Hicksville, N.Y.: Exposition Press, 1975.

Human Relations Area Files. Area Handbook: Saudi Arabia. Washington, D.C.: Human Relations Area Files, 1958.

Hunter, Robert. The Dilemma of Soviet Involvement in the Middle East. London: Institute of Strategic Studies, 1969/1970.

Hurewitz, J. C. Changing Military Perspectives in the Middle East. Santa Monica, Calif.: Rand, 1971.

__________. Diplomacy in the Near and Middle East. 2 vols. Princeton: D. Van Nostrand, 1956.

__________. Middle East Politics: The Military Dimension. New York: Frederick A. Praeger, 1969.

__________, ed. Soviet-American Rivalry in the Middle East. New York: Frederick A. Praeger, 1969.

Ickes, Harold L. Fighting Oil. New York: Alfred A. Knopf, 1943.

Inglis, K. A. D., ed. Energy: From Surplus to Scarcity? New York: John Wiley & Sons, 1974

Insight Team of the Sunday Times. Insight on the Middle East War. London: Andre Deutsch, 1974.

Iskander, Marwan. The Arab Boycott of Israel. Beirut: Research Center, Palestine Liberation Organization, 1966.

Ismael, Tareq Y. The Middle East in World Politics: A Study in Contemporary International Relations. Syracuse: Syracuse University Press, 1974.

Ismail, Salem K. The Correlation Between Energy Consumption and Gross National Product. Vienna: Organization of Petroleum Exporting Countries, 1968.

Issawi, Charles, and Mohammed Yeganeh. The Economics of Middle Eastern Oil. New York: Frederick A. Praeger, 1962.

_________. Oil, the Middle East and the Third World. New York: The Library Press, 1972.

Jabber, Fuad. Israel and Nuclear Weapons: Present Options and Future Strategies. London: Chatto and Windus, 1971.

Jacoby, Neil H. Multinational Oil: A Study in Industrial Dynamics. New York: Macmillan Co., 1974.

Jensen, W. G. Energy and the Economy of Nations. Oxfordshire: G. T. Fouls & Co., 1970.

Karpat, Kemal H., ed. Political and Social Thought in the Contemporary Middle East. New York: Frederick A. Praeger, 1968.

Keesing's Research Report. The Arab-Israeli Conflict: The 1967 Campaign. New York: Charles Scribner's Sons, 1968.

Kenen, I. L. et al. A Just Peace in the Mid-East: How Can It Be Achieved? Washington, D.C.: American Enterprise Institute for Public Policy Research, 1971.

Kerr, Malcolm H. The Arab Cold War 1958-1970. London: Oxford University Press, 1971.

_________. The Middle East Conflict. New York: Foreign Policy Association, 1968.

Key, Kherim K. The Arabian Gulf States Today. Washington, D.C.: Asia Research Center, 1974.

Khadduri, Majid. Arab Contemporaries: The Role of Personalities in Politics. Baltimore: Johns Hopkins University Press, 1973.

_________. Political Trends in the Arab World. Baltimore: Johns Hopkins University Press, 1972.

Khouri, Fred J. The Arab-Israeli Dilemma. Syracuse: Syracuse University Press, 1968.

Kirk, George. Contemporary Arab Politics. New York: Frederick A. Praeger, 1961.

________. Survey of International Affairs: The Middle East in the War 1939-1946. London: Oxford University Press, 1952.

Klebanoff, Shoshana. Middle East Oil and U.S. Foreign Policy. New York: Praeger, 1974.

Kohler, Foy D. et al. The Soviet Union and the October 1973 Middle East War; The Implications for Detente. Miami: Center for Advanced International Studies, University of Miami, 1974.

Laqueur, Walter Z. Communism and Nationalism in the Middle East. New York: Frederick A. Praeger, 1956.

________. Confrontation: The Middle East and World Politics. New York: Bantam Books, 1974.

________, ed. The Israel-Arab Reader: A Documentary History of the Middle East Conflict. New York: Bantam Books, 1970.

________. The Struggle for the Middle East: The Soviet Union and the Middle East 1958-70. Baltimore: Penguin, 1972.

Lebkicher, Roy et al. Aramco Handbook. The Netherlands: Arabian American Oil Company, 1960.

________. Aramco and World Oil. New York: Arabian American Oil Co., 1952.

Leeman, Wayne A. The Price of Middle East Oil: An Essay in Political Economy. Ithaca, N.Y.: Cornell University Press, 1962.

Lenczowski, George. The Middle East in World Affairs. Ithaca, N.Y.: Cornell University Press, 1968.

________. Oil and State in the Middle East. Ithaca, N.Y.: Cornell University Press, 1960.

________. Russia and the West in Iran, 1918-1948: A Study in Big Power Rivalry. Ithaca, N.Y.: Cornell University Press, 1949.

________. Soviet Advances in the Middle East. Washington, D.C.: American Enterprise Institute for Public Policy Research, 1971.

________, ed. The Political Awakening in the Middle East. Englewood Cliffs, N.J.: Prentice Hall, 1970.

_________, ed. United States Interests in the Middle East. Washington, D.C.: American Enterprise Institute for Public Policy Research, 1968.

Lewis, Bernard. The Arabs in History. New York: Harper & Row, 1966.

_________. The Middle East and the West. New York: Harper & Row, 1964.

Lilienthal, Alfred M. The other Side of the Coin: An American Perspective of the Arab-Israeli Conflict. New York: Devin-Adair, 1965.

Lipsky, George A. Saudi Arabia: Its People, Its Society, Its Culture. New Haven, Conn.: HRAF (Human Relations Area Files) Press, 1959.

London, Kurt, ed. New Nations in a Divided World: The International Relations of the Afro-Asian States. New York: Frederick A. Praeger, 1965.

Long, David E. Middle East Problem Paper No. 10: Confrontation and Cooperation in the Gulf. Washington, D.C.: Middle East Institute, 1974.

Longhurst, Henry. Adventure in Oil. London: Sidgwick and Jackson, 1959.

Longrigg, Stephen H. The Middle East: A Social Geography. Chicago: Aldine. Atherton, 1963.

_________. Oil in the Middle East: Its Discovery and Development. London: Oxford University Press, 1961.

Lubell, Harold. Middle East Oil Crisis and Western Europe's Energy Supply. Baltimore: Johns Hopkins University Press, 1963.

Lufti, Ashraf. Arab Oil: A Plan for the Future. Beirut: Middle East Research and Publishing Center, 1960.

_________. OPEC Oil. Middle East Oil Monographs. No. 6. Beirut: Middle East Research and Publishing Center, 1968.

Macrakis, Michael S., ed. Energy: Demand, Conservation, and Institutional Problems. Cambridge, Mass.: MIT Press, 1974.

Makdisi, Samir A. Middle East Problem Paper No. 12: Oil Price Increases and the World Economy. Washington, D.C.: Middle East Institute, 1974.

Malvestiti, Piero. Sources of Energy and Industrial Revolutions. Publications Departments of the European Communities, 1961.

Mancke, Richard. The Failure of U.S. Energy Policy. New York: Columbia University Press, 1974.

McCracken, Paul W. Is the Energy Crisis Contrived? Washington, D.C.: American Enterprise Institute for Public Policy Research, 1974.

McDonald, Stephen L. Petroleum Conservation in the United States: An Economic Analysis. Baltimore: Johns Hopkins University Press, 1971.

McLane, Charles B. Soviet-Middle East Relations. London: Central Asian Research Center, 1973.

McLean, John G., and Robert Wm. Haigh. The Growth of Integrated Oil Companies. Boston: Harvard University, 1954.

McNamara, Robert S. The Essence of Security: Reflections in Office. New York: Harper & Row, 1968.

Medvin, Norman. The Energy Cartel: Who Runs the American Oil Industry. New York: Vintage Books, 1974.

Mikdashi, Zuhayr M. The Community of Oil Exporting Countries. Ithaca, N.Y.: Cornell University Press, 1972.

________. A Financial Analysis of Middle Eastern Oil Concessions: 1901-65. New York: Frederick A. Praeger, 1966.

________, and Hollis B. Chenery. Arabian Oil. Chapel Hill: University of North Carolina Press, 1949.

Mikesell, Raymond F. et al. Foreign Investment in the Petroleum and Mineral Industries. Baltimore: Johns Hopkins University Press, 1971.

Miller, Roger LeRoy. The Economics of Energy: What Went Wrong and How We Can Fix It. New York: William Morrow and Co., Inc., 1974.

MIT Energy Laboratory Policy Study Group. Energy Self-Sufficiency: An Economic Evaluation. Washington, D.C.: American Enterprise Institute for Public Policy Research, 1974.

Mitchell, Edward J., ed. Dialogue on World Oil. Washington, D.C.: American Enterprise Institute for Public Policy Research, 1974.

Mitchell, Edward J. U.S. Energy Policy: A Primer. Washington, D.C.: American Enterprise Institute for Public Policy Research, 1974.

Monroe, Elizabeth. The Changing Balance of Power in the Persian Gulf. New York: American Universities Field Staff, 1972.

_________, and Robert Mabro. Oil Producers and Consumers: Conflict or Cooperation. New York: American Universities Field Staff, 1974.

Mosley, Leonard. Power Play: Oil in the Middle East. Baltimore: Penguin Books, 1974.

Mughraby, Muammad A. Permanent Sovereignty Over Oil Resources. Beirut: Middle East Research and Publishing Center, 1966.

Nash, Gerald D. United States Oil Policy 1890-1964. Pittsburgh: University of Pittsburgh Press, 1968.

Nasrollah, S. Fatemi. Oil Diplomacy: Powderkeg in Iran. New York: Whittier Books, 1954.

National Petroleum Council. Impact of New Technology on the U.S. Petroleum Industry, 1946-1965. Washington, D.C.: National Petroleum Council, 1967.

_________. Petroleum Resources Under the Ocean Floor. Washington, D.C.: National Petroleum Council, 1969.

_________. U.S. Energy Outlook. Washington, D.C.: National Petroleum Council, 1972.

Nutting, Anthony. The Arabs. New York: New American Library, 1964.

O'Connor, Harvey. World Crisis in Oil. New York: Monthly Review Press, 1962.

Odell, Peter R. An Economic Geography of Oil. New York: Frederick A. Praeger, 1963.

_________. Oil and World Power: A Geographical Interpretation. London: Cox & Wymen, 1972.

Odum, Howard T. Environment, Power, and Society. New York: Wiley, 1971.

Pachachi, Nadim. The Role of OPEC in the Emergence of New Patterns in Government-Company Relationships. London: Royal Institute of International Affairs, 1972.

Penrose, Edith T. The Growth of Firms, Middle East Oil and Other Essays. London: Frank Cass, 1971.

__________. The Large International Firm in Developing Countries: The International Petroleum Industry. London: George Allen and Unwin, 1968.

Pennar, Jaan. The U.S.S.R. and the Arabs: The Ideological Dimension. New York: Crane, Russak, 1973.

Peretz, Don. Israel and the Palestine Arabs. Washington, D.C.: Middle East Institute, 1958.

__________. The Middle East Today. New York: Holt, Rinehart and Winston, 1965.

Philby, H. St. John B. Arabia. New York: Scribner, 1930.

__________. Arabian Jubilee. New York: John Day Co., 1953.

__________. The Empty Quarter. London: Constable, 1933.

Pranger, Robert J. American Policy for Peace in the Middle East 1969-1971. Washington, D.C.: American Enterprise Institute for Public Policy Research, 1971.

__________, and Dale R. Tahtinen. Nuclear Threat in the Middle East. Washington, D.C.: American Enterprise Institute for Public Policy Research, 1975.

Pringgle, Laurence. Energy: Power for People. New York: Macmillan Co., 1975.

Quandt, William B. et al. The Politics of Palestinian Nationalism. Berkeley: University of California Press, 1973

Ramazani, Rouhollah K. The Northern Tier: Afghanistan, Iran, and Turkey. Princeton, N.J.: D. Van Nostrand Co., 1966.

__________. The Persian Gulf: Iran's Role. Charlottesville: University of Virginia Press, 1973.

Ridgeway, James. The Last Play: The Struggle to Monopolize the World's Energy Resources. New York: E. P. Dutton & Co., 1973.

Rifai, Taki. The Pricing of Crude Oil: Economic and Strategic Guidelines for an International Energy Policy. New York: Praeger, 1974.

Rocks, Lawrence, and Richard P. Runyon. The Energy Crisis. New York: Crown Publishers, 1972.

Roosevelt, Kermit. Arabs, Oil and History: The Story of the Middle East. New York: Harper and Row, 1949.

Rouhani, Fuad. A History of O.P.E.C. New York: Praeger, 1971.

Roumaini, Maurice M., ed. Forces of Change in the Middle East. Worcester, Mass.: Worcester State College Press, 1971.

Rustow, Dankwart A. Middle Eastern Political Systems. Englewood Cliffs, N.J.: Prentice-Hall, 1971.

__________. Politics and Westernization in the Near East. Princeton, N.J.: Center of International Studies, 1956.

Safran, Nadav. From War to War: The Arab-Israeli Confrontation 1948-1967. New York: Pegasus, 1969.

__________. The United States and Israel. Cambridge: Harvard University Press, 1963.

Said, Abdul A., ed. America's World Role in the 70s. Englewood Cliffs, N.J.: Prentice-Hall, 1970.

__________, and Luiz R. Simmons, eds. The New Sovereigns: Multinational Corporations as World Powers. Englewood Cliffs, N.J.: Prentice-Hall, 1975.

Sayegh, Fayez A. Arab Unity: Hope and Fulfillment. New York: Devin-Adair Co., 1958.

Sayyeq, Kamal S. Oil and Arab Regional Development. New York: Frederick A. Praeger, 1968.

Schmidt, Dana Adams. Armageddon in the Middle East. New York: John Day Co., 1974.

Schurr, Sam H. et al. Middle Eastern Oil and the Western World: Prospects and Problems. New York: American Elsevier Publishing Co., 1971.

__________. Energy in the American Economy 1850-1975: An Economic Study of Its History and Prospects. Baltimore: Johns Hopkins University Press, 1960.

Sharabi, H. B. The Government and Politics of the Middle East in the Twentieth Century. New York: D. Van Nostrand, 1962.

_________. Nationalism and Revolution in the Arab World. Princeton, N.J.: D. Van Nostrand, 1966.

Shwadran, Benjamin. The Middle East, Oil and the Great Powers. New York: Frederick A. Praeger, 1955.

Sinai, I. Robert et al. Modernization and the Middle East. New York: American Academic Association for Peace in the Middle East, 1970.

Spiegel, Steven L., ed. At Issue: Politics in the World Arena. New York: St. Martin's Press, 1973.

Stocking, George W. Middle East Oil: A Study in Political and Economic Controversy. Kingsport, Tenn.: Vanderbilt University Press, 1970.

Stork, Joe. Middle East Oil and the Energy Crisis. New York: Monthly Review Press, 1975.

Tahtinen, Dale R. The Arab-Israeli Military Balance Today. Washington, D.C.: American Enterprise Institute for Public Policy Research, 1973.

_________. Arms in the Persian Gulf. Washington, D.C.: American Enterprise Institute for Public Policy Research, 1974.

Tanzer, Michael. Energy Crisis: World Struggle for Power and Wealth. New York: Monthly Review Press, 1974.

_________. The Political Economy of International Oil and the Underdeveloped Countries. Boston: Beacon Press, 1969.

Taylor, Alan R. Prelude to Israel: An Analysis of Zionist Diplomacy 1897-1947. Beirut: Institute for Palestine Studies, 1970.

Thomson, J. H., and R. D. Reischauer. Modernization of the Arab World. New York: D. Van Nostrand, 1966.

Toynbee, Arnold J. Survey of International Affairs, Vol. 1, 1925: The Islamic World. London: Oxford University Press, 1927.

Tugendhat, Christopher. Oil: The Biggest Business. New York: G. P. Putnam's Sons, 1968.

Twitchell, K. S. Saudi Arabia. Princeton, N.J.: Princeton University Press, 1958.

Vicker, Ray. The Kingdom of Oil. New York: Charles Scribner's Sons, 1974.

Walpole, Norman C. et al. Area Handbook for Saudi Arabia. Washington, D.C.: American University Press, 1971.

Ward, Barbara. The Rich Nations and the Poor Nations. New York: W. W. Norton & Co., 1962.

Whetten, Lawrence L. The Soviet Presence in the Eastern Mediterranean. New York: National Strategy Information Center, 1971

Yager, Joseph A., and Eleanor B. Steinberg. Energy and U.S. Foreign Policy. Cambridge, Mass.: Ballinger Publishing Co., 1974.

Yannacone, Victor J., ed. Energy Crisis: Danger & Opportunity. St. Paul: West Publishing Co., 1974.

GOVERNMENT PUBLICATIONS

Abdo, Albert N. Saudi Arabia. Washington, D.C.: U.S. Government Printing Office, 1962.

Department of State. Saudi Arabia: Background Notes. Washington, D.C.: U.S. Government Printing Office, May 1973.

Embassy of Saudi Arabia. Prince Faisal Speaks. Washington, D.C., 1962.

International Economic Report of the President. Washington, D.C.: U.S. Government Printing Office, February 1974.

Ministry of Information, Kingdom of Saudi Arabia. Faisal Speaks. Book 12, n.d.

The Budget of the United States Government, Fiscal Year 1976 (Appendix). Washington, D.C.: Government Printing Office, 1975.

Saudi Arabia Monetary Agency. Annual Report. Jeddah, 1972 and 1973.

U.S. Congress, House. Balance of Payments Adjustment to Higher Oil Prices: Managing the Petrodollar Problem/Report of the Ad Hoc Committee on Domestic and International Monetary Effect of Energy and Other Natural Resource Pricing of the Committee on Banking and Currency. 93rd Cong., 2nd sess. Washington, D.C.: U.S. Government Printing Office, 1975.

__________. Developing Countries and the United States in the World Economy: Problems and Prospects/Report of the Ad Hoc Committee on the Domestic and International Monetary Effect of Energy and Other Natural Resource Pricing of the Committee on Banking and Currency. 93rd Cong. 2d sess. Washington, D.C.: U.S. Government Printing Office, 1974.

__________. Energy Security and the Domestic Economy: Impact on Prices, Employment, and Consumption/Report of the Ad Hoc Committee on the Domestic and International Monetary Effect of Energy and Other Natural Resource Pricing of the Committee on Banking and Currency. 93rd Cong. 2d sess. Washington, D.C.: U.S. Government Printing Office, 1973.

__________. Hearings before the Subcommittee on the Near East and South Asia of the Committee on Foreign Affairs. 93rd Cong. 1st sess. Washington, D.C.: U.S. Government Printing Office, 1973.

INTERNATIONAL ORGANIZATION PUBLICATIONS

International Bank for Reconstruction and Development, and International Development Association. Additional External Capital Requirements of Developing Countries. Washington, D.C., March 1974.

International Monetary Fund. Memorandum. Washington, D.C., November 4, 1974.

Organization for Economic Cooperation and Development, Special Committee for Oil. Oil Today. Paris, 1964.

United Nations. Review of Economic Conditions in the Middle East. E/1910 Add. 2/Rev. 1. New York, 1951.

United Nations, Department of Economic and Social Affairs. Petroleum Exploration: Capital Requirements and Methods of Financing. New York, 1962.

__________. Utilization of Oil Shale: Progress and Prospects. New York, 1967.

United Nations General Assembly Doc. A/PV. 1555. New York, 1967.

CONFERENCE AND SYMPOSIUM PUBLICATIONS

Akins, James E. "Evolving Relationships Among the Oil Companies, the Oil Producing Governments: Confrontation or Cooperation?" World Energy Demands and the Middle East. Washington, D.C.: The Middle East Institute, 1972.

Lichtblau, John H. "Arab Oil and a Settlement of the Middle East Conflict." In After the Settlement: New Directions, New Relationships. Papers delivered at the 28th Annual Conference of the Middle East Institute. Washington, D.C., October 12, 1974.

Simon, Douglas W., and Richard S. Rhone. "The Future of the United Nations in a Changing International System." Paper delivered at the annual meeting of the New Jersey Political Science Association, Seton Hall University, April 20, 1974.

Summary Proceedings of a Symposium on World Food and Population. Washington, D.C.: American University, 1974.

Yamani, Shaykh Ahmad Zaki, "Prospects for Cooperation Between Oil Producers, Marketers and Consumers: The Issue of Participation and After," World Energy Demands and the Middle East. Papers delivered at the 26th Annual Conference of the Middle East Institute, Washington, D.C., September 1972.

ARTICLES

Adelman, Morris A. "Is the Oil Shortage Real? Oil Companies as OPEC Tax-Collectors." Foreign Policy, Winter 1972-73.

__________. "Oil Prices in the Long Run 1963-75." Journal of Business (University of Chicago), April 1964.

Akins, James. "The Oil Crisis: This Time the Wolf is Here." Foreign Affairs 51, no. 3 (April 1973).

Amuzegar, Jahangir. "The Oil Story: Facts, Fiction and Fair Play." Foreign Affairs 51, no. 4 (July 1973).

Avery, Peter, "Iran's Foreign Policy." New Middle East, no. 47 (August 1972).

Balfour-Paul, H. G. "Recent Developments in the Persian Gulf." Royal Central Asian Journal (London) 56, pt. 1, (February 1969).

Ball, George W. "The Looming War in the Middle East and How to Avert It." Atlantic, January 1975.

Barger, Thomas C. "Middle Eastern Oil Since the Second World War." Annals, May 1972.

Battle, Lucius D. "Peace-Inshallah." Foreign Policy, no. 14, (Spring 1974).

Berry, John A. "Oil and Soviet Policy in the Middle East." Middle East Journal, Spring 1972.

Best, Geoffrey. "Middle East Oil and the U.S. Energy Crisis: Prospects for New Ventures in a Changed Market." Law and Policy in International Business 5, no. 1 (1973).

Bill, James A. "The Challenge of Chance: Petroleum and Planning in the Middle East." Focus 22, no. 1 (September 1971).

Brechenfeld, Gurney. "How the Arabs Changed the Oil Business." Fortune, August 1971.

Campbell, John C. "The Arab-Israeli Conflict: An American Policy." Foreign Affairs 49, no. 1 (October 1970).

Carter, William G. "National Support of Multinational Ventures." Columbia Journal of World Business 7, no. 5 (September-October 1972).

Chenery, Hollis B. "Restructuring the World Economy." Foreign Affairs 53, no. 2 (January 1975).

Cottrell, Alvin J. "British Withdrawal from the Persian Gulf." Military Review 50, no. 6 (1970).

_________. "Conflict in the Persian Gulf." Military Review 51, no. 2 (1971).

_________. "Iran, the Arabs and the Persian Gulf." Orbis, Fall 1973.

_________. "A New Persian Hegemony." Interplay 3, no. 12 (September 1970).

_________. "Shah of Iran Concerned over Saudi Arabia's Future." New Middle East, no. 31 (1971).

_________. "The U.S. and the Future of the Gulf after the Bahrain Agreement." New Middle East, no. 22 (1970).

Dammann, T. "Saudi Arabia's Dilemma—An Interview with King Faisal." Interplay 3, no. 12 (September 1970).

Demaree, Allan T. "Arab Wealth As Seen Through Arab Eyes." Fortune, February 1974.

__________. "Aramco Is a Lesson in the Management of Chaos." Fortune, February 1974.

Desjardins, T. "Saudi Arabia: Next in Line for Revolution?" Atlas 19, no. 9 (September 1970).

Edens, David G., and William P. Snavely. "Planning for Economic Development in Saudi Arabia." Middle East Journal, Winter 1970.

Eilts, Hermann, "Social Revolution in Saudi Arabia, Part 1." Parameters 1, no. 1 (1971).

Farmanfarmaian, Khodadad et al. "How Can the World Afford OPEC Oil." Foreign Affairs 53, no. 2 (January 1975).

Field, Michael. "Oil: OPEC and Participation." World Today 28, no. 1 (January 1972).

Gardner, Frank J. "Participation Pact Signals Power Shift." Oil and Gas Journal, October 16, 1972.

__________. "Russian Oil Exports Head for Squeeze." Oil and Gas Journal, October 20, 1969.

Gaspard, J. "Faisal's Arabian Alternative." New Middle East, no. 6 (March 1969).

Hartshorn, J. "Oil Diplomacy: The New Approach." World Today 29, no. 7 (July 1973).

Hottinger, Arnold. "The Depth of Arab Radicalism." Foreign Affairs 51, no. 3 (April 1973).

Hunter, Robert E. "The Energy Crisis and U.S. Foreign Policy." Headline Series, no. 216, Foreign Policy Association, June 1973.

__________. "The Soviet Dilemma in the Middle East, Part II: Oil and the Persian Gulf." Adelphi Papers (London), Institute for Strategic Studies, October 1969.

Ingotus, Miles. "Seizing Arab Oil: A Blueprint for Fast and Effective Action." Harper's, March 1975.

Issawi, Charles. "Oil, the Middle East and the World." The Washington Papers 4, Center for Strategic and International Studies, Washington, D.C., 1972.

Kelly, John B. "The Future in Arabia." International Affairs 42, no. 4 (October 1966).

Kirk, George. "Ibn Sa'ud Builds an Empire." Current History, December 1934.

Knauerhase, Ramon. "Saudi Arabia's Economy at the Beginning of the 1970s." Middle East Journal 28, no. 2 (1974).

Knowles, Ruth Sheldon. "A New Soviet Thrust, the Arab-Israeli Crisis and Two Big Iraq Agreements Help the Russian's Major Oil Putsch in the Middle East." Middle East, December 1969.

Kraar, Louis. "The Shah Drives to Build a New Persian Empire." Fortune, October 1974.

Laird, Melvin R. "Let's Meet the Energy Crunch—Now." Reader's Digest, January 1975.

Levy, Walter J. "World Oil Cooperation or International Chaos." Foreign Affairs 52, no. 4 (July 1974).

Lewis, John P. "Oil, Other Scarcities, and the Poor Countries." World Politics 22, no. 1 (October 1974).

Lof, George O. G. "Solar Energy: An Infinite Source of Clean Energy." Annals, November 1973.

Luce, Charles F. "A Battle Plan to Beat the Energy Crisis." Reader's Digest, March 1975.

Luttwork, Edward N., and Walter Laqueur. "Kissinger and the Yom Kippur War." Commentary 58, no. 3 (September 1974).

Mabro, R., and Elizabeth Monroe. "Arab Wealth from Oil: Problems of its Investment." International Affairs 50, no. 1 (January 1974).

Magnus, Ralph H. "Middle East Oil." Current History, February 1975.

McCarthy, Terence. "The Middle East: Will We Go to War." Ramparts, March 1975.

McKie J. "The Political Economy of World Petroleum." American Economic Review 64, no. 2 (May 1974).

Melkus, R. "Toward a Rational Future Energy Policy." Natural Resources Journal 14, no. 2 (April 1974).

Peek, Malcolm C. "Saudi Arabia's Wealth: A Double-Edged Sword." New Middle East, no. 40 (January 1972).

Penrose, Edith. "Origins and Development of the International Oil Crisis." Millenium 3, no. 1 (1974).

Pollack, Gerald A. "The Economic Consequences of the Energy Crisis." Foreign Affairs 52, no. 3 (April 1974).

Ramey, James T. "The Promise of Nuclear Energy." Annals, November 1973.

Rentz, George. "Saudi Arabia: The Islamic Island." Journal of International Affairs 19, no. 1 (January 1965).

Rondot, Pierre. "Palestine: Peace Talks and Military." World Today 30, no. 9 (September 1974).

Rose, David J. "Energy Policy in the U.S." Scientific American, January 1974.

__________. "Nuclear Electric Power." Science 184, no. 4134 (April 19, 1974).

Rostow, Eugene V. "A Basis for Peace." The New Republic, April 5, 1975.

Rouleau, Eric. "The Palestinian Quest." Foreign Affairs 53, no. 2 (January 1975).

Rugh, William. "Emergence of a New Middle Class in Saudi Arabia." Middle East Journal 27, no. 1 (1973).

Safran, Nadav. "Engagement in the Middle East." Foreign Affairs 53, no. 1 (October 1974).

__________. "The War and the Future of the Arab-Israeli Conflict." Foreign Affairs 52, no. 1 (October 1974).

Sayigh, Yusif. "Problems and Prospects of Development in the Arabian Peninsula." International Journal of Middle East Studies 2, no. 1 (January 1971).

Schurr, Sam H. "Energy." Scientific American, September 1973.

Sheehan, Edward R. F. "Mastermind of Mideast Oil." Reader's Digest, September 1974.

Smith, Ian. "The Super-Powers and the Middle East." World Today, January 1974.

Stauffer, Thomas R. "Oil Money and World Money: Conflict or Confluence?" Science 184, no. 4134 (April 19, 1974).

Taylor, Alan. "The Isolation of Israel." Journal of Palestine Studies 4, no. 1 (Autumn 1974).

Turner, Louis. "Politics of the Energy Crisis." International Affairs 50, no. 3 (July 1974).

Ullman, Richard H. "After Rabat: Middle East Risks and American Roles." Foreign Affairs 53, no. 2 (1975).

Varin, Bension, and Kenj Takeuchi. "Developing Countries and Non-Fuel Minerals." Foreign Affairs 52, no. 3 (April 1974).

Waverman, Leonard. "Oil and the Distribution of International Power." International Journal 29, no. 4 (August 1974).

Wolf, Martin. "Solar Energy Utilization by Physical Methods." Science 184, no. 4134 (April 19, 1974).

Wilson, Carroll L. "A Plan for Energy Independence." Foreign Affairs 51, no. 4 (July 1973).

Yamani, Sheikh Ahmed Zaki, "Oil: Towards a New Producer-Consumer Relationship." World Today, November 1974.

PERIODICALS

The periodicals listed below are exclusive of those already cited in articles above.

Arab Observer (Cairo), August 27, 1962 and October 1, 1962.

Arab Oil and Gas Journal (Beirut), March, May 1971; May 1974.

Business Week, January 13, 1975.

Economist (London), November 3, 1973.

Far Eastern Economic Review (Hongkong), December 19, 1963.

Fortune, May 1975.

Journal of Palestine Studies, Autumn 1973.

Middle East Economic Digest (London), November 7, 1969; December 7, 28, 1973; March 15, 22, July 19, 1974.

Middle East Economic Survey (Beirut), November 18, 1966; July 21, 1967; June 7, September 13, 1968; August 11, 1972; June 15, 1973.

Middle East Record, 1967.

New York Times Magazine, March 24, 1974.

Newsweek, April 9, May 21, November 12, December 10, 1973; February 4, 18, March 28, July 1, September 30, October 7, 14, 28, December 16, 23, 30, 1974; January 13, 1975.

Oil and Gas Journal, April 17, December 25, 1972; August 6, October 22, 1973; March 24, 1975.

Oil Forum, April 1952.

Organization for Economic Cooperation and Development, Press Release, November 15, 1974.

Petroleum Press Service (London), January, November 1973.

Science, June 15, 1973.

Scientific American, September 1971.

Time, October 14, November 4, 25, 1974; January 6, April 7, 1975.

Transcript of "Face the Nation," CBS Television Network, April 20, 1974.

U.S. News & World Report, January 15, 1973; December 30, 1974; and March 10, 1975.

Washingtonian, March 1975.

NEWSPAPERS

Christian Science Monitor, July 6, September 4, 1973; November 18, 1974.

Daily Telegraph (London), November 17, 1973.

New York Times, December 26, 1963; June 21, 1972; October 18, December 31, 1973; September 19, 24, October 13, 30, November 16, 19, December 17, 1974; February 20, May 5, 1975.

Sunday Times (London), November 14, 1973.

Times (London), November 9, 1917.

Washington Post, September 11, 1962; June 17, October 17, December 24, 26, 31, 1973; February 5, 14, March 28, April 6, May 13, 26, June 9, 15, 16, August 25, September 24, 30, October 5, 6, 14, 24, 30, November 6, 16, 23, December 1, 7, 10, 14, 16, 22, 30, 1974; January 3, 12, February 4, 20, 24, March 1, 28, April 2, 15, 19, May 23, June 5, 1975.

Washington Star-News, December 13, 18, 1974.

INDEX

ABOUT THE AUTHOR

SHEIKH RUSTUM ALI is currently a foreign service officer of Bangladesh assigned to its Embassy in Washington, D. C. He is the first Bangladesh diplomat to obtain a Ph. D. degree, and his dissertation, on which this study is based, was the first on petrodiplomacy completed at The American University in the summer of 1975. Dr. Ali is one of the few specialists on petrolism from the developing countries. He serves as a consultant to the Institute for Third World Studies in Manchester, United Kingdom.

Dr. Ali was an Assistant Professor of International Relations at the University of Dacca in Bangladesh before coming to the United States. He holds triple masters degrees in Government, Political Science, and International Relations from New York University, the University of the Philippines, and the University of Dacca, respectively. He was a Fulbright scholar at New York University. Dr. Ali, a noted scholar, petrologist, and diplomat, who has travelled extensively in the Middle East, South Asia, Far East, Europe, and North America, is uniquely qualified to give a non-partisan interpretation of the Arab oil diplomacy.

Date Due